The Science of Cooking. A Quick Immersion

Quick Immersions uses accurate and straightforward language to offer a good introduction, or deeper knowledge, on diverse issues, as well-structured texts by prestigious authors delve into the worlds of political and social sciences, philosophy, science and the humanities.

Claudi Mans

THE SCIENCE OF COOKING
A Quick Immersion

Tibidabo Publishing

Copyright © 2018 by Claudi Mans.
Published by Tibidabo Publishing, Inc.

All rights reserved. No part of this publication may be reproduced, stored in a retrieval system, or transmitted, in any form or by any means, electronic, mechanical, photocopying, recording, scanning or otherwise, without the prior permission in writing of the Published, or as expressly permitted by law, or under terms agreed with the appropriate reprographics rights organization.

Copyediting by Lori Gerson
Cover art by Raimon Guirado
For illustrations, tables and graphics credits, please see page 7.

First published 2019

Visit our Series on our Web:
www.quickimmersions.com

ISBN: 978-1-949845-06-8
1 2 3 4 5 6 7 8 9 10

Printed in the United States of America.

Contents

List of illustrations, tables and graphics 7
Introduction. Start by Frying Croquettes 9
1 A Quick Immersion: Why Cook? 13
2. A Bit of Food Chemistry 23
   2.1 Bromatology and Food Chemistry 25
   2.2 Labels on Prepared Foods 29
   2.3 Range of Food Products. Objects and Substances 35
   2.4 Purity. Solutions and Mixtures 41
   2.5 Acids, Bases and Salts in the Kitchen 49
3. Molecules, the Actors Responsible for
Food's Organoleptic Properties 53
   3.1 The Color of Foods 54
   3.2 Smell, Taste and Flavor in Foods 70
4. Biological Structures. Where are the Nutrients? 82
5. Food Texture 87
   5.1 Liquids: Solutions 90
   5.2 Solids 92
   5.3 Texturizers 95
6. Dispersed Systems, the Key to Many
Culinary Preparations 97
7. Physical Culinary Processes 107
   7.1 Mechanical Processes 109
   7.2 Thermal Processes above Room Temperature 111
   7.3 Thermal Processes at Low Temperatures 116

8. Chemical or Physicochemical Culinary Processes — 117
  8.1 Denaturation of Proteins, a Physicochemical Process — 119
  8.2 Fermentations — 123
  8.3 Caramelization — 132
  8.4 Maillard Reactions — 134
  8.5 Spherifications — 137
  8.6 Other Chemical Reactions in the Kitchen: Undesired Browning — 140
  8.7 Aging — 142

9. Three Kitchen Utensils: A Comparison — 143
  9.1 Pots, Pressure Cookers and Vacuum Cookers — 144
  9.2 The Frying Pan and the Deep Fryer — 152
  9.3 The Oven and the Microwave Oven — 157

10. More Devices and Methods — 162
  10.1 Coffee Makers — 162
  10.2 Cooking with Liquid Nitrogen — 165
  10.3 Distillation in the Kitchen — 166
  10.4 Lyophilization — 167

11. Are There Limits in the Kitchen? — 169

12. FAQs — 173
  Further Reading — 179

List of illustrations, tables and graphics

Figure 1. Outline of basic processes in a kitchen — 21
© Claudi Mans

Figure 2. Various food coloring molecules — 57
© Claudi Mans

Figure 3. Dispersed systems — 106
© Claudi Mans

Figure 4. Outline of basic spherification — 139
© Claudi Mans

Table 1. Overlapping terminology between science and cooking — 15 to 17
© Claudi Mans/Alicia & el Bullitaller

Table 2. Triglycerides and fats — 44
© Claudi Mans

Table 3. Food coloring before being processed — 60 to 62
© Claudi Mans

Table 4. Principal sweeteners classified as food additives by order of INS number — 73-74
© Claudi Mans

Table 5. Some emulsifiers, thickeners and gelling agents, classified by order of their INS number — 101
© Claudi Mans

Picture 1. Food label 34
© Petrnutil (petrnutil/123RF.COM)

Picture 2. Chocolate. Various textures
of the same foodstuff 88
© Anton Starikov (coprid/123RF.COM)
© Belchonock (belchonock/123RF.COM)
© Oleksandr Farion (farion25/123RF.COM)

Picture 3. Beer fermentation process 127
© Татьяна Кузьмичева (Tatyanakyz/123RF.COM)

Picture 4. Kitchen utensils 144
© Viktoriia Tarnopolska (vitarpan/123RF.COM)

Picture 5. Molecular cuisine. Cheesecake
with chocolate cooked in liquid nitrogen 165
© Alexey Kostin (Kostin77/123RF.COM)

Picture 6. Lyophilization.
Fresh and freeze-dried orange 168
© Viktoriia Adamchuk (missv2506/123RF.COM)

Introduction.
Start by Frying Croquettes

The destiny of food is to feed, for which it must be prepared, cooked, savored, digested and absorbed by the organism. Food technologists prepare food so it can reach our mouths, and dieticians, nutritionists and doctors worry about what happens to the food inside of us, and what happens inside of us from the food. The interface between these two worlds, the exterior and the interior, is the mouth and its surrounding area. It is the chefs and experts in gastronomic sciences who take care of this part. And this is the setting for this book: how chemistry provides knowledge and technology for the preparation of nutritious, tasty and healthy food, which is why we focus on the kitchen. We will take a short excursion both outwards and inwards, but will attempt to stay in our area.

If the reader wishes to prepare some chicken croquettes for dinner, they can follow various methods: buy frozen croquettes already battered and ready to fry, buy prepared croquettes at their butcher's shop or delicatessen, or take the time to prepare the croquette dough and then batter and fry them. We are going to follow this last option, the most difficult, to take advantage of leftovers. We will look for a recipe online that seems reliable and do it as best we can, improvising on what cannot be followed exactly.

We take some leftover roasted chicken along with some sautéed vegetables, such as peppers or onion, and chop it all up with a knife or a meat grinder. We will grind the chicken until the chunks are the size we want for the croquette. Now we will make a bechamel sauce by taking some butter, which we will heat in a saucepan, adding in flour until the ingredients become very gooey. Then we add milk in small quantities, stirring it well so it mixes with the dough. When we have enough quantity and its texture is correct, very thick, we mix the bechamel with the chicken and chopped vegetables, along with some salt and, if we want, some grated nutmeg. We shape the croquettes into the desired size and coat them in breadcrumbs and beaten egg.

Then we need to fry them, and here there are also three basic options: we can use a frying pan with a finger's worth of hot oil, or a deep fryer with a few cups of oil. Alternatively, we can prepare them in the oven, practically without oil. If we use a deep fryer, which is the option I recommend, we set it at 330-340°F and when it is ready, we immerse the croquettes in the oil until they are golden brown. We then place them on a paper towel and they are now ready to eat.

All of the above is **the what**, what all cooks should do to prepare a dish. But what the reader wishes to know is **the why**, given that they have bought a book about science in the kitchen. Throughout the process of making the croquettes we can ask ourselves, for example, why frozen croquettes are usually smaller than the ones from the butcher's shop or our own. Or why we give them an oval shape and not a round one. Or what happens to the flour when we mix it with butter so that it turns dark. Or why we batter them with breadcrumbs and beaten egg, and do not just fry them directly. Or why the temperature should be 330ºF not 300ºF. Or how many times we can use the oil in the deep fryer. These many questions have different levels of depth –geometric, chemical, physical, technological– that require knowing the 'whys' on different descriptive levels. These different levels and sciences involved are what make the kitchen a complex world, whose explanation and understanding require a certain order.

Responding to the whys requires knowing details of the ingredients we are using and the particularities of the processes. Specifically, and for our example, let's go to a level that delves deeper than the previous questions. We must know how the chicken is cooked, and why its skin ends up golden and crunchy and the meat not so much. Why the vegetables get soft when they are cooked. What are the particles of flour like, and what happens to the flour when it is mixed with the melted butter and, before this, what is butter. And what is bechamel, and why does it have properties between an elastic solid and a more or less gooey liquid. And, also, what reactions are responsible for the golden crust on the croquette or on the chicken skin. And how does the heat from the coat of oil manage to penetrate inside the croquette. And what are the reactions that change the composition of the oil in the deep fryer over time and the number and type of foods fried in it.

All this leads us to the need to know three basic and fundamental aspects: their **composition**, that is, what chemical substances the foods are made of and what gives them color, aroma and flavor; and their **structure**, that is to say, how are these substances arranged, forming the animal and vegetable biological structures –cells, organs and organisms– that are the ones we will use in the kitchen. And, then, **how** composition and structure is **modified** when the food is subjected to changes in temperature, alone or in the presence of other ingredients: the physical changes and the chemical reactions. All of this will allow us to understand, with greater or lesser depth, how each culinary technique works, and even in what direction culinary novelties are heading in the future in regard to both techniques and ingredients. As such, knowledge on what we have and judicious use of science allows us, in a certain way, to predict the future: it is the recipe for everything in life, and in the kitchen as well.

Chapter 1
A Quick Immersion: Why Cook?

We have seen that there is more than one basic technique to fry something. The frying we used for the croquettes is a type of **deep frying** that consists in rapidly submerging the food in hot oil and keeping it there long enough for it to be sufficiently fried. "Deep frying" can mean two things: that the food is in contact with the hot oil for a short period of time, to avoid that it burns too much; or the immersion is done all at once, brusquely, and not slowly, like how some people enter a chilly swimming pool when the water is too cold for them.

This book is a quick immersion in the second sense of these meanings. From the first line of the introduction we stepped inside the subject, directly and without complaints such as "chemistry is an underrated science etc. etc." But it is not a quick immersion in the sense of lasting a short time: it is not a brief dip and then let's lay in the sun. We cover the subjects with the necessary length of time, respecting the space the editor has given us; and the final objective of the process is that something has changed in the reader's mind upon finishing the book. And this something should be to assume that procedures in the kitchen and cooking are chemical and physical processes of transforming the original raw material into something safer, healthier and more appetizing thanks to methods that we humans have been domesticating over years, centuries and millennium; that these methods can be satisfactorily explained by the sciences; and that these same sciences that explain what happens to food can help chefs and food technologists to invent new preparations in the future that are better for our nutrition and satisfaction.

Language and **terminology** will be an important part of this book. Special emphasis will be placed on the overlapping terminology between chemistry and cooking, highlighting the cases in which the term has different meanings in each of the settings (Table 1). For example, the word *sugar* in cooking language means table sugar, that is, the sweet deriving from beets or sugarcane, while in chemistry the term *sugar* includes a wide range of substances with similar characteristics, such as sucrose –table sugar– as well as glucose, lactose, levulose, maltose and other similar substances, some of which are sweet and others are not. But sucralose, with the same suffix in its name, is not a sugar and does not come from the same chemical family, nor does cellulose. Conversely, diose and psicose are indeed sugars, although they do not seem to be, etc. (Alicia & elBullitaller, 2006).

acid   in science: substance capable of donating H^+ ions in an aqueous solution (classical definition, not current). Has a pH of less than 7.
in cooking: sour, tart.

aerosol   in science: suspension of solid or liquid particles in the air.
in cooking: pressurized container used to make things frothy or foamy.

alkali   in science: substance capable of accepting H^+ ions in an aqueous solution (classical definition, not current).
in cooking: what neutralizes an acid.

base   in science: alkali.
in cooking: any preparation that is basic for a dish.

cook   in science: transform foods at a temperature higher than room temperature.
in cooking: transform foods in any way and at any temperature.

dense   in science: having high mass by unit of volume.
in cooking: thick, gooey.

emulsion   in science: system of dispersing drops of an aqueous liquid in another that is oily, or vice versa.
in cooking: any preparation that inextricably mixes different types of liquids and solid particles until getting a very homogenous mixture that does not separate when left to sit.

foam   in science: dispersion system formed by gas bubbles separated by liquid membranes.
in cooking: gelling agent in which gas bubbles are suspended; can be whipped or from a canister.

gel   in science: fine particles dispersed in a continuous medium in which the liquid medium has become viscous enough to behave more or less like a solid.
in cooking: food with a gelatinous texture.

milk   in science: secretion from female mammals to feed their offspring.
in cooking: milk from mammals used in cooking, or milk made from any type of crushed vegetable seed and mixed with water.

oil   in science: triglyceride derived from fatty acids.
in cooking: any fat that is liquid at room temperature.

pure   in science: having no other substances in its composition.
in cooking: having not been subject to alterations originally or during preparation.

salt	in science: neutral substance formed by anions and cations.
	in cooking: table salt, chiefly sodium chloride (NaCl).
sugar	in science: any sweet-tasting carbohydrate that has 6 to 12 carbon atoms in its molecule.
	in cooking: sucrose, table sugar having the formula $C_{12}H_{22}O_{11}$.
thick	in science: synonym for dense.
	in cooking: synonym for gooey.

Table 1. Overlapping terminology between science and cooking. (Alicia & elBullitaller, 2006).

As this paragraph is being written there are more than 145 million described chemical substances, and a good part of these are in organisms in nature. It would not be feasible, nor of much interest, to write long lists of which substances are in every type of food. The discourse needs to be structured so that it is not only of interest, but also useful to provide a comprehensive perspective. In truth, to talk about science in the kitchen requires starting by looking at what foods are, from a chemical point of view; that is, what chemical substances they are formed of. But it is necessary to immediately see the relationship between their composition and the biological structures that make up the foods. In the following pages we will verify that the substances and structures that constitute foods give them a good part of the sensory properties we look for in them: color, aroma, flavor and texture, as well as their nutritional properties. We will see later how sensory and nutritional properties can be modified by means of cooking techniques, especially when cooking at various temperatures, and how we can take advantage of our knowledge to get the desired characteristics in our culinary preparations. This is the book's outline, but the reader should be warned that its development will not be as linear as the previous paragraph might suggest. And that is because everything is connected. As of now the author asks forgiveness for the inevitable repetitions the reader will find in the book, which efforts have been made to reduce.

The term to cook refers to two different procedures. Generically, it refers to all the processes that are done in the kitchen and, more specifically, to those processes in which the food is transformed when heated.

Living beings are –we are– organisms open to the environment, with which we exchange matter, energy and information. Food constitutes the matter we receive in the broadest sense, and waste is also sent to the environment. We also receive energy in different forms, and we send waste energy externally in the form of heat. This happens in every cell, in every grouping of cells called an organ, in every grouping of organs called an organism, and in the groupings of organisms in their medium, called ecosystems.

Humans are organisms and we have to feed ourselves. In addition to the gas we absorb to breathe –air is the nourishment we ingest in the greatest quantity, some 22 or 26 pounds per day– we feed on water and other liquids, and on various solids. According to our metabolism, foods give us three types of substances: **structural** substances, to grow and continually replace our cells that are dying; **energy** substances, which fundamentally provide us with the chemical energy contained in their structures and allow us to maintain our organism, overall, and each organ in particular, alive; and **regulatory** substances that help our metabolism to function correctly.

The Components of Food

The substances contained in foods are usually classified according to their chemical composition in **carbohydrates**, basically to supply energy; **lipids**, also called fats, also basically for energy; and **proteins**, mainly structural and also to provide energy; and vitamins and mineral salts, basically regulatory. Fiber is also usually added to these components, although it is not absorbed by our organism and functions only as a facilitator for digestion. And water, almost omnipresent in foods. All our liquid and solid foods contain these substances, in a very wide assortment of forms. In packaged foods, the quantity of each one is listed on the label.

Figure 1 summarizes the processes in any kitchen. The primary producers are farmers, ranchers and those who fish. They supply their unprocessed products to chefs, the food industry or to private individuals. All these people then transform the items into processed products that, after being mixed with other ingredients, some perhaps provided by industrial companies, constitute what is prepared. These are the bases for the dishes that are finally served. From the perspective of this book, what is important is that in each of the transformations the ultimate culinary objective, from the gastronomic point of view, takes place: to modify the structure and physical and chemical characteristics of the products, changing their aroma, their color, their flavor, their texture or their presentation.

Ferran Adrià, in his provocative talks, usually distinguishes between a strawberry picked directly off a bush and eaten, which is not to cook, and this same strawberry nicely presented on a plate, perhaps accompanied by something, which is indeed to cook, although there is no type of cooking. **Cooking means to subject a food to all those processes that allow it to be fed to an eater**: peeling, chopping, cutting, mixing, grinding, whipping, preparing emulsions, salting, freezing, etc., and cooking.

It is also **cooking foods by means of using a temperature that is higher than room temperature**. There are many ways to cook: boil, fry, sauté, brown, bake, sear and a thousand other operations, all of them at temperatures noticeably above room temperature. Many chefs also consider that certain operations such as preparing fish immersed in cold acid baths –ceviche, anchovies in vinegar– are also cooking, and that it is even cooking when something is quickly frozen by immersing it in liquid nitrogen at -320ºF, as will be discussed in Chapter 9.

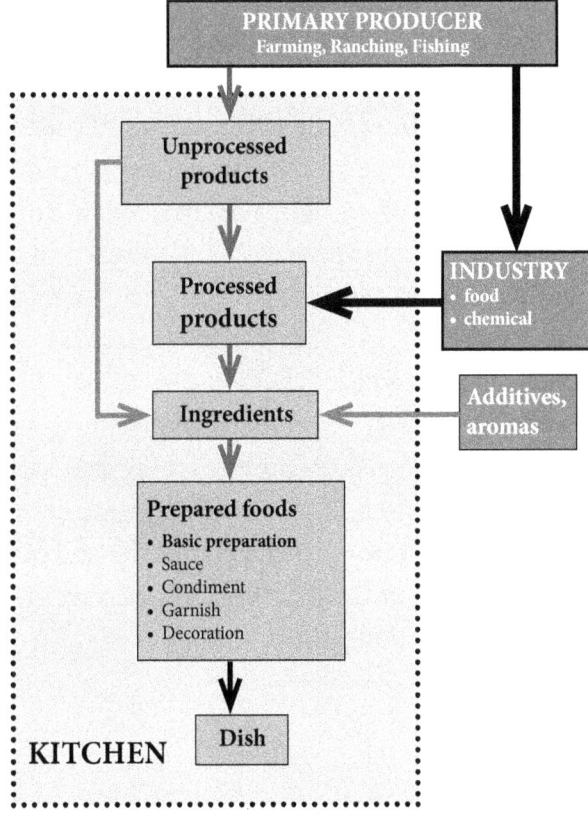

Figure 1. Outline of basic processes in a kitchen. Preparing a dish involves the primary producer who supplies the raw materials, the industry that may provide some ingredients, and the chef who prepares the dishes.

Cooking at temperatures significantly above room temperature have different functions. It is sufficient to cite three of them here, all of which will be developed in more detail in later chapters. Proteins are denatured at temperatures above 160ºF: the egg white of a hard-boiled egg is an example of denatured proteins in gel form. Pathogenic organisms are destroyed because their organism is denatured for the same reason, and as such the cooked food is safer. At higher temperatures fats melt, collagen deteriorates and starches gelatinize. All of this changes the texture and improves palatability of the foods, permitting substances that were undigestible before to be digested, like what happens with potatoes, wheat and certain hard parts of animals. And at temperatures above 300ºF you get the Maillard reactions and caramelizations, which change the color, flavor and aroma of foods, in general making them more appetizing. Cooking, then, provides more food safety, facilitates digestion and improves the gastronomic aspects of food. The reference noted at the end of the book (This, H. *et al*, 2017) covers a good part of the subject of science and gastronomy.

Chapter 2
A Bit of Food Chemistry

The science of chemistry contributes different tools to be able to understand what the ingredients and products we use in the kitchen are, how they change with different culinary processes and, an aspiration still not completely achieved, how we should treat ingredients and substances to get a desired preparation. For this reason, chemists use chemical analysis techniques to know what there is in a substance and techniques of a structural character, such as different microscopes, to understand the structure of substances. Because, as has been said and will be repeated on other occasions, it is not only the composition that is relevant in the science of cooking, but also –above all– the intimate structure of the food.

There is an enormous number of edible plant species, with thousands of varieties. For example, it is calculated that there has been and are more than 10,000 types of tomatoes between those that are cultivated and those that grow wild. And the same for all types and species of plants and animals. In recent years, globalization has permitted species that were once exotic to be commercialized in far-flung markets. Farming techniques have allowed varieties that were nearly extinct to be brought back, and biotechnological techniques –technologies applied to the study, design and modification of human beings in all senses– have raised expectations to create varieties that do not exist in nature. **Transgenic varieties** (known as **GMO, g**enetically **m**odified **o**rganisms) of plants and animals even get genes from one species put into the genome of another, which gives it some advantages. More recently (2013), the CRISPR-Cas9 technique allows the gene sequence to be changed in a single organism, which in certain cases produces resistance to bacterial attacks, among other characteristics. Biotechnological techniques are debated socially, but have been used for decades to grow soybeans and corn. All experts agree on the safety of transgenic products, to be consumed by people as well as animals, and also for the environment. This is why they are authorized in many countries. However, in Europe the use of transgenic products is prohibited to feed humans, but not to feed animals, due to the influence of anti-transgenic campaigns on public opinion. In the countries where they are allowed, such as the USA, there are campaigns asking that products containing these be labelled.

The enormous variety of foods bring with them the enormous variety of chemical substances they are comprised of, of which there are millions. But fortunately, from the perspective of chemistry in the kitchen, as well as from nutritionists and dieticians, the large majority of these substances can be grouped in the categories already mentioned earlier: carbohydrates, fats, proteins, minerals, vitamins, fiber and water. Caviar and a sardine seem to be unrelated, but both have fats, proteins and water, and the fats they contain have similar dietary properties, although clearly they are not the same from a gastronomic point of view. It is outside the scope of this book to describe, even without details, the composition of the different foods. For this there are very comprehensive books (McGee, 2007), and here just a little is going to be said. Instead, general processes applicable to the majority of foods will be described, but they will not be described as cooking recipes, because they are not.

2.1. Bromatology and Food Chemistry

When did chemists begin to sniff around in kitchens? Ever since chemistry is chemistry –starting with

Robert Boyle and his book *The Sceptical Chymist* in 1661– culinary phenomena and chemical transformations in the kitchen have interested scientists, but obviously before that there had also been important contributions to what happens in the kitchen. In a technological perspective, in 1679 **Denis Papin** invented the Marmite, a pressure cooker, to take advantage of residues from the meat industry and turn them into edible jellies. It was probably **Justus von Liebig**, around 1840, who had a more complete scientific view of the world of chemistry applied to food, with an already modern perspective. His studies on the chemistry of fertilizers led him to understand that certain nutrients control plant growth: especially phosphate, nitrogen and carbon dioxide salts. And his studies on nutrition in plants, animals and humans provided him with evidence that all organisms can synthesize proteins, fats and complex sugars from their foods, even if these organisms do not contain them. This is the equivalent of accepting that chemical reactions, which make some substances appear and others disappear, are carried out inside any living organism. In addition to being a scientist, Liebig had a very applied view of his knowledge, which led him to patent the meat extract that carries his name, and also baking powder.

Liebig's global studies and, later, those of **Louis Pasteur** and other scientists allowed the development of **bromatology**, a science that studies food as a whole, from its production to its assimilation by the organism. Throughout the 20th and 21st centuries, advances in bromatology, along with those in physics and chemistry and different technologies, have allowed the food industry to develop, have notably advanced knowledge about nutrition, science-based cuisine has developed and, unfortunately, the number of diets and myths about food, having no scientific basis, have grown enormously.

> ### Science and Cooking
>
> **Nicholas Kurti** confirmed in 1969 that *"the temperature inside the stars is better known than the temperature inside a souffle."* This is when the current movement linking science and cooking began, which had as its pioneers **Kurti**, a physicist, and **Hervé This**, a chemist, and those who have followed an entire generation of scientists interested in culinary phenomena, and another generation of chefs interested in the science hidden inside their preparations. The result has been the creation of a movement called molecular **gastronomy**, a name not to everyone's liking —I include myself among those who do not like it— but has earned a fortune and become popular among large swaths of the professional and non-professional population. Proof of that is the book you have, reader, in your hands.

A crucial part of bromatology is **food chemistry**, that is, knowledge about their composition, their properties and their behavior when processed. If we ask ourselves what the composition of a substance is, there is not only one answer, as it can be approached on different levels. If what you want to know is which of the 188 known elements form part of its composition, the chemical tool to use is **elemental analysis**. By means of different techniques, the proportion of each chemical element in the substance can be known: in the majority carbon, oxygen, nitrogen, hydrogen, and the rest of the elements in lesser proportion: sulfur, phosphorous, sodium, magnesium, iron or heavy metals. Those in lesser proportions are called **trace elements**, and they are decisive in metabolic processes.

In contrast, if what you want to know are the molecules and other chemical entities present in a substance, the tools to use are the extremely varied techniques in analytical chemistry, many of them using sophisticated instruments. For most substances of food-related interest, complete identification of all their components is an extremely complex and labor-intensive analysis because the number and variety of molecules present are very high. The branch of chemistry that identifies components of food and all types of biological substances is **organic chemistry of natural products**. This science is constantly discovering new molecules of food-related and pharmaceutical interest in living organisms, including the human body; it identifies them, characterizes them and determines their chemical structure, and studies their possible applications.

Various of these molecules will appear, with their chemical names, throughout this book. The complexity of many of them means that on most occasions only their names will be given, but not their chemical formulas nor their structure. The interested reader can easily find more information on the internet. Their structures include carbon and hydrogen atoms and, depending on the product, also oxygen, nitrogen, sulfur or phosphorous, and many other elements, in structures with hundreds or thousands of atoms intertwined in complex tridimensional molecules.

2.2. Labels on Prepared Foods

The labels on mineral water are an example of elemental analysis combined with the identification of some simple chemical species: it is possible to read on them the amounts of cations from metals such as sodium, potassium or magnesium, along with the amounts of some anions such as bicarbonate, sulfate, silicates and others.

A radically different perspective to the previous is the nutritional characterization of a food, that is to say, predicting the effect its ingestion will have on the organism. In many cases it is enough to know what molecules it contains, classified by type, but an exhaustive list of them is not necessary. The

nutritional information of a certain commercial product is reflected on its label. The amount of each one of the groups of basic nutrients already mentioned are shown here: carbohydrates, fats, proteins, fiber, minerals, vitamins, salt –which is usually specifically indicated although it is in the mineral group– and other minor components. The nutritional information is completed with the energy value of the food, which should be given in units of energy from the International System of Units, the *Joule* (J), a name derived from the Scottish physicist **James Joule**. For historical reasons, *calories* (cal) and *kilocalories* (kcal, with 1 kcal = 4180 J) also continue to be used. Each one of the nutrient families, for example fats, usually contains many chemical species, each one with a molecule different from the others, but all similar. They can all be grouped together because from a nutritional perspective all the chemical species of a certain family are equivalent. As such, a pound of any sugar has an energy value of 1812 kcal, the same as proteins, and any type of fat has an energy value of 4000 kcal/lb.

The **list of ingredients** of a prepared food is obligatory information on a label, and is very useful for the consumer. Appearing on this list are the substances the creator has used to prepare the product, but chemical substances do not appear on these lists as they are not ingredients, except in the case of food additives. Details on all the molecules present in a food are not usually useful, apart from the fact that in many cases they are not well-known. For example, the indication that a food contains olive oil is enough from the point of view of the consumer, but is not very specific from the chemical perspective because olive oil –independent of its purity and quality– is a natural mixture of many different but very similar compounds, all of them derived from different fatty acids, among which predominates oleic acid. The list of ingredients on the labels, in accordance with legal criteria, includes from more to less all the ingredients used in its creation. Some of them are pure substances from the chemical viewpoint: water, sugar. Other ingredients are complex mixtures, such as milk, or derivatives from animals or plants such as wheat flour, or anatomical parts from organisms such as ham, and even complete biological species such as garlic or tomato.

Another set of techniques exists that helps with the chemical and structural characterization of a substance. Due to the fact that the majority of foods derive from living organisms, *cellular biology* and its microscopic approach with the entire range of available instruments –optical microscope, electronic microscope, scanning tunneling microscope and others– helps us to characterize the cells and their components. At the same time, techniques from *biochemistry* and *molecular biology* provide knowledge on which molecules make up each one of the cell organelles, where and how they are arranged and how they transmit the information from one end of the organism to the other. In this way, a fairly complete outline currently exists that goes from the molecule to the cell, and from this to the organism. This knowledge is basic in current medicine and also in other branches of knowledge such as gastronomy and the sensory sciences: for example, because of this we know which molecules are involved in the sense of smell and how they interact on the molecular level with the receptors present in the pituitary.

Many other analytical viewpoints fit under this chapter's title, some of which fall completely outside the scope of this book. Everything –everything material– can be described in chemical terms, and the detailed chemical description of foods could become an endless and unintelligible list. At times, this leads to extrapolations that are not sufficiently precise. Take, for example, the approximate chemical composition of an apple taken from the internet. Its ingredients would be (I transcribe literally): Alpha-Linolenic-Acid, Asparagine, D-Categin, Isoqurctrin, Hyperoside, Ferulic-Acid, Farnesene, Neoxathin, Phosphatidyl-Choline, Reynoutrin, Sinapic-Acid, Caffeic-Acid, Chlorogenic-Acid, P-Hydroxy-Benzoic-Acid, P-Coumaric-Acid, Avicularin, Lutein, Quercitin, Rutin, Ursolic-Acid, Protocatechuic-Acid, Silver, Tryptophan, Threonine, Isoleucine, Leucine, Lycine, Methionine, Cystine, Phenylalanine, Tyrosine, Valine, Argenine, Histidine, Alanine, Aspartic Acid, Glutamic Acid, Glycine, Proline, and Serine. Trace amounts of Boron, Cobalt, Proteins, Insoluble Fiber, Water, Carbohydrates, Lipids, Saturated fats, traces of pesticides and fertilizers (don't worry they are TRACES after all) and many many more that haven't been discovered at all. This type of description –well-intentioned, of course– seeks to demonstrate that "everything is chemical," and that, as such, there is no need to be afraid of chemistry, nor of chemical language or chemical products. But such an approach presents various inconveniences and, in this specific case, also diverse errors. Their enumeration and criticism can be useful for the purposes of this book.

- First error: They are not **ingredients**. This term can only be used for prepared mixtures, such as packaged foods, or in kitchen recipes. An apple, legally, does not have ingredients because nobody that is not nature –a very relative nature, as it deals with agriculture– has made it. It would be better to talk about components.
- Second error: It lists **compounds** and **elements** without distinguishing between them. In fact, there is water or asparagine in the apple, but not cobalt or silver as elements but rather their compounds.
- Third error: Product **families** are included on the list simultaneously with individual products from the family itself or its components. For example, the protein family is composed of many amino acids on the list, such as tryptophan.
- Fourth error: It does not specify important compounds present in the apple, but rather summarizes this in generic "carbohydrates."

- Fifth error, which for me is the most important: The list seems to reduce the chemistry of the apple to the mixture of a group of substances, as if the apple were a sack full of substances which, on their own, explained all the properties of the apple. But what determines many of its properties is not only the chemical substances but how they are structured.

Nutrition Facts

Serving Size
Serving per Container

Amount Per Serving

Calories	Calories From Fat	
	% Daily Value *	
Total Fat	...g	...%
Saturated Fat	...g	...%
Cholesterol	...g	...%
Sodium	...mg	...%
Total Carbohydrate	...g	...%
Dietary Fiber	...g	...%
Sugar	...g	...%
Protein	...g	...%

Vitamin A	...%	•	Vitamin C	...%
Calcium	...%	•	Iron	...%

*Percent Daily Values are based on a 2,000 calorie diet. Your daily values may be higher or lower depending on your calorie needs.

Picture 1. Food label.

Effectively, the texture of the food, how it is cooked, its conservation and even its nutritional properties are explained not only by its composition but, above all, by the form the aforementioned compounds have in the fruit. That is, what is the *macroscopic* structure of the fruit –epicarp, mesocarp, endocarp, kernel– and what is its *microscopic* structure. That is to say, what are the plant cells of each of these parts like, and where and in what context are the chemical compounds mentioned located: Are they in a solution in the cytoplasm? Do they form part of some cell structure, such as the membrane? Or are they in the middle of an extracellular liquid? An apple, in the context of this book, is an *object*. In view of this, it is important to have a wide perspective when speaking about chemistry in the kitchen, given that it does not only have to do with chemistry, but also biology and physics.

2.3. Range of Food Products. Objects and Substances

All foods come from nature, principally the biosphere, whether plants, animals or microorganisms. Very few foods come from the mineral world: only water, salt and some minerals used for nutritional therapies. And in much lower quantities, dozens of synthetic food additives have their origins in mineral fuels such as coal or oil. The treatment initial materials undergo until becoming ingestible foods can be considerably complex, and the food that is finally consumed can be very far from its natural origin due to the transformations it has experienced. The geographic origin could also be very distant, but this is an aspect that does not concern us here, even if in the mentality of many consumers the preference is increasing to consume foods produced nearby, called zero food miles, slow food, local production or lifecycle analysis. The concept of *nearness* here is more an idea of marketing than of geography. In the food industry, and by extension in the rest of the sectors involved in food, foods are usually classified into five –and even six– food products, depending on the treatment they have received.

- **Primary food products** have not undergone any type of cooking or preparation beyond some initial pre-treatment to make them easier to package or transport. Fruits and vegetables, uncooked meats and fish, milk and eggs are primary food products. Also included in these products are salted foods, dehydrated products and those that are fermented.
- **Secondary food products** are conserved and preserved foods, in cans or metallic or glass bottles that have been subject to sterilization. The foods have been chopped, cooked, mixed with other ingredients and canned. Examples of these are cans of tuna, marmalades or fruits in syrup.
- The **tertiary food products** are commonly called processed foods. These are ready-to eat, or heat-and serve foods, such as TV dinners and airline meals. They have generally been cooked, then are packaged and kept cold. For their consumption they need to be heated in a conventional oven or a microwave. Frozen pizzas and frozen trays of meat or fish are examples of these products.

On the other hand, in Europe the classifications are somewhat different. The primary and secondary products coincide, but the **tertiary food products** are quick-frozen and frozen foods. Examples of these are fruits, vegetables, fish and shellfish with no further processing other than their cleaning, chopping and freezing. They should remain in their frozen state (at least -4ºF) during their entire storage and transport time. This is known colloquially as *maintaining the cold chain*.

- The **fourth food products** are those that are pre-made and packaged in a protective atmosphere. Some examples would be salads, packaged fruit that is cut and peeled, peeled and pre-cut potatoes and similar products, still without final cooking. They are stored in cold and should also maintain the cold chain between the temperature they are frozen at and the storage temperature in the refrigerator, which is around 40ºF.

In Europe, the **fifth food products** are those described previously as tertiary food products.

There is no agreement among experts on if there is an additional **food product**, which in Europe would be the sixth. For some these would be foods conserved by means of irradiation. For others, the sixth food products would be those that are freeze-dried, such as food for astronauts or for expeditions, or pre-cooked noodles that you only need to add water to increase their volume and reconstitute the food. The future will specify in what direction the industry and the consumer go.

> ### Varieties of Kitchens
>
> There are many types of kitchens depending on the purpose of their activity and their production scale.
>
> **Food company kitchens** are industrial production plants similar to giant kitchens, and they prepare all types of food products, from secondary to fifth, always packaged, for sale to distributors.
>
> **Kitchens at institutions** such as hospitals, company cafeterias, school kitchens or airplane catering prepare all the range of prepared food products according to their characteristics and final recipient. As such, depending on the case, they prepare standard dishes, children's portions, menus for diabetics, for people with dysphagia or difficulties swallowing, for those who are lactose intolerant, coeliacs, menus without allergens, kosher or halal dishes, or dishes for vegans. The menus for people on airplanes must be heated in the air, and are good examples of fifth food products.
>
> **Gastronomic restaurant kitchens** in general try to use primary food products as their raw materials, and at the same time some launch their own fifth food product dishes for marketing and sale.
>
> The final consumer in their **home kitchen** uses primary to fifth food products, depending on their food ideology, family needs, available time,

> economic possibilities and family habits. Many home kitchens in the developed world are currently under-exploited, as the option of meals shared simultaneously by all members of the family is losing followers, who prefer to eat at a restaurant or eat pre-cooked dishes, perhaps made at home, but consumed away from the family household.

Let's now take a look at another concept to classify foods. A very elemental classification in any kitchen would be to distinguish between objects and substances, already suggested earlier. All whole animals and plants, or their pieces and cuts, are **objects**: a partridge, a radish, a chicken thigh, an apple, an egg. While they are not processed, the objects keep a defined form. In the classic elemental classification of states of matter –solid, liquid, gas– the objects correspond to solids. All the objects mentioned derive directly from nature through the activities of agriculture, livestock breeding or fishing, but there are also many other objects invented by man and which come from an industrial or domestic process. Examples of this are a noodle, a croquette, a square of chocolate or a sausage. In the kitchen, objects are processed, in principle, as individualities.

Substances, on the contrary, are sold in bulk, by mass or volume, but not by units: there are liquid substances, such as milk or wine, and powdered substances derived from solids, such as flour, sugar, powdered chocolate or pollen. These last substances are made up of solid particles of very small sizes, and in certain quantities the grouping has some of the properties of liquids, such as adapting to the shape of the container, or flowing more or less easily when poured. Grains of cereals or legumes do not adapt as well to this classification: the smaller each unit is, the more characteristics of "substance" and less characteristics of "object" they have; quinoa grains have a diameter of 1 mm or less, and they flow easily. On the contrary, fava beans are, one by one, objects, although in practice they are cooked in mass.

This crude classification may appear to be a mere anecdote, but it has importance in culinary processes. Objects, especially if they have some centimeters of dimension, are usually cooked by applying heat from the outside –except in microwave ovens– and as such their surface is heating while the interior is still at room temperature. Depending on the dimensions, the degree of cooking desired, and on the properties of the matter the food is made of, the cooking should be more or less slow, and at a higher or lower temperature. Conversely, in mixtures of substances, especially if they are dispersed in a liquid, in many cases they can be considered to be homogenous mixtures, and that all the mass is at the same temperature. In this case, the distinction between the temperatures on the surface and inside the mass is usually irrelevant.

2.4. Purity. Solutions and Mixtures

The concept of **purity** in chemistry is very strict, because specialists in chemical analysis over time invent ever more sophisticated devices that allow impurities to be detected –*traces*, in technical language– that before went unnoticed. Substances can be detected in concentrations of *1 ppt*, which means one part per trillion, that is, 1 in 1,000,000,000,000. It is the equivalent of detecting one person hidden among all the inhabitants of one hundred planets like today's Earth. Conversely, the criteria for purity in culinary practice is much less restrictive.

Pure Substances

What **pure substances** do we have in the kitchen?

- **Water**, for practical purposes, can be considered a pure substance, both from the tap or mineral water, and has only H_2O molecules, disregarding the small amount of minerals it might contain. The composition of salts in mineral waters is quite varied, from those with very weak mineralization to those that are more saline. In any case, the amount of salts in more saline mineral waters is around 0.19lb./cu.ft., which in water indicates a purity of 99.7%. Waters with weak mineralization have a water purity of 99.95% or more. Sparkling water contains significative amounts of natural or added carbon dioxide.

- **Sugar**. The term *sugar* in colloquial language indicates the carbohydrate derived from beets or sugarcane. White sugar has a purity of 99.9% sucrose. It is appropriate to point out that arguments demonizing refined sugar for its composition are absurd because brown sugar and all its varieties also contain, in very high proportions, white sugar, up to 98%. A different question is that, surely, the different brown sugars that exist have some minimum additional nutritional properties owing to the small proportion of molasses they contain.
- **Salt**. The term salt in everyday language indicates common salt or table salt. Conversely, in chemical terms a salt is any compound formed by cations and anions. When salt is refined it can have a richness of 99.5% sodium chloride. Common table salt has a richness from 95 to 98%, and the remainder are other salts found in the sea, such as sulfate and magnesium chloride. *Sea salt* is of a similar composition, somewhat less rich in sodium chloride. Other salts with more or less bright colors, such as *Himalayan salt* –which is obtained in Pakistan, some hundreds of miles from the Himalayas– contains a certain amount of clays and gypsum that give it a reddish or pink tint. These impurities do not give it any appreciable nutritional value. It has 98% sodium chloride, like all salts. Its advertising states that it contains up to 84 different elements, which is chemically absurd.

- Liquid **nitrogen**. This liquid is of course not usual in household kitchens, but is indeed in certain restaurant kitchens and in the frozen food industry. This is the chemical element nitrogen, the main component in air, and it can remain liquid at -320ºF at atmospheric pressure. It separates from air through techniques at very low temperatures and is a substance of high purity because, for certain uses, it is of interest that the oxygen content be very low to avoid oxidation.
- **Food additives**, used in the prepared food industry, are very pure substances, almost all with a known molecular structure. For example, additive 300 (INS number) is the antioxidant ascorbic acid or Vitamin C, and 960 the sweetener *steviol glycoside*. It is colloquially known as *stevia*, although this is really the name of the plant it is extracted from and not the chemical name of its active ingredient.

Simple Mixtures

The most common **simple mixtures** in the kitchen are usually liquids:

- **Oil**. This name applies to many types of substances from different chemical families that includes mineral fuel oils –diesel oil– lubricating oils and edible oils. Focusing on these last ones, oils come from three sources: oils coming from fruits –olive, a paradigm for oils, or avocado; seed oils –sunflower, canola, grapeseed, peanut, corn, walnut; and those coming from animals, such as tallow, whale oil, lard or milk fat. All oils and fats are basically mixtures of triglycerides (Table 2).

A	$C_3H_8O_3$ CH_2-OH | $CH-OH$ | CH_2-OH	Two ways of representing glycerin or glycerol
B	$C_{18}H_{34}O_2$	Three ways of representing oleic acid
	$COOH-(CH_2)_7-CH=CH-(CH_2)_7-CH_3$	
	$COOH-CH_2-CH_2-CH_2-CH_2-CH_2$ $-CH_2-CH_2-CH=CH--CH_2-CH_2-$ $CH_2-CH_2-CH_2-CH_2-CH_2-CH_3$	
C	$COOH-(CH_2)_{16}-CH_3$	Stearic acid
	$COOH-(CH_2)_7-CH=CH-CH_2-CH=$ $CH-(CH_2)_4-CH_3$	Linoleic acid
D	$CH_2-O-CO-(CH_2)_7-CH=$ $CH-(CH_2)_7-CH_3$ | $CH-O-CO-(CH_2)_7-CH=$ $CH-(CH_2)_7-CH_3$ | $CH_2-O-CO-(CH_2)_7-CH=$ $CH-(CH_2)_7-CH_3$	Triglyceride from oleic acid

Table 2: Triglycerides and fats. A: Formulas condensed and developed from glycerin or glycerol. B: Formulas condensed, semi-developed and developed from oleic acid, monounsaturated as it has a double bond in its chain. C: Formulas semi-developed from stearic acid, saturated (it has no double bond in its chain), and from linoleic acid, polyunsaturated as it has two double bonds in its chain. D: Formula semi-developed for triglyceride from oleic acid.

Triglycerides are complex molecules that animals and plants synthesize, coming from glycerin and fatty acids. These are carbon chains with various ways of linking, which allows distinguishing between saturated, monounsaturated and polyunsaturated fatty acids, some of which are the well-known *omega-3* and *omega-6*, and others the *trans*. All animal and plant species generate oils or fats with a mixture of triglycerides, of which some are dominant. For example, the dominant triglyceride in olive oil is the one coming from *oleic acid*. All triglycerides are less dense than water and insoluble in it, whether transparent or translucent, and can mix with each other, which at times has given rise to commercial adulterations. When olive oils are obtained by simple cold pressing they are cloudier because they contain some substances in suspension coming from the olive pulp. Given that oils are mixtures of different molecules, they are not *chemically pure* substances. That does not mean the oil is not pure. The concept of purity has a different meaning in food preparation.

- **Wine** is a mixture obtained from the fermentation of grape or other fruit must. It contains water, ethyl alcohol or ethanol, natural coloring, residual sugar from the must (mainly glucose), glycerin, other alcohols and many more substances. Some of these do not exist in the original grape and come from the ethanol fermentation. Certain additives are allowed in wine, among them sulfites to avoid its oxidation.

In spite of the high number of components in wine, we give it the qualification of a simple mixture because it is simply a hydroalcoholic solution.
- **Vinegar** is also a mixture, derived from wine by later oxidation through acetic fermentation. It is a mixture with many components, especially water and acetic acid which, like wine, is chemically a solution.
- **Liquors** and **spirits** are, in the great majority, hydroalcoholic mixtures with or without sugar, whose color and flavor derive from the extraction or distillation of flavors from different plants. All of them are simple mixtures from the chemical point of view, even though dozens or hundreds of plants may have intervened in their preparation, as asserted by Jägermeister (56 plants) or Chartreuse (130 plants).

Complex Mixtures

Complex mixtures are classified as such because their internal structure is not the same as a simple solution, but rather they are comprised of various mixed phases or states –solids, liquids and gases. These are **dispersed systems**, which are called this because they are made up of at least two immiscible substances, that is, ones not capable of combining to make a heterogenous mixture, for example oil and water, in which one of them is dispersed in small drops in the middle of the other liquid. Among these complex mixtures we find natural products such as milk and prepared products like sauces or ice creams (see Chapter 6).

- **Milk** from any mammal is a dispersed system having natural origins. This is a solution of water with lactose in which drops of milk fat and calcium caseinate particles can be found, which in advertising is what is called the *calcium in milk*. This structure in which drops from a liquid phase are dispersed in the middle of another liquid phase is called **emulsion**. The emulsion of oil-in-water is usually abbreviated as O/W (*oil in water*).
- **Butter** is also an emulsion, but in this case the fat is the continuous medium and it has small drops of water with lactose dispersed in the middle. It is then a water-in-oil emulsion or W/O *(water in oil)*. Butter is usually kept refrigerated, which keeps it from being clearly seen as an emulsion because the fat acts as a solid, but indeed the emulsion is noted when it is melted and is sufficiently heated: the drops of water boil and you can see the vapor in the form of foam on the surface of the melted butter.
- **Sauces** are also usually dispersed systems. **Mayonnaise**, for example, is an emulsion in which the continuous medium is the water from the egg white and from the yolk, with the proteins from the egg white, while the fat from the yolk and oil are dispersed in the form of drops.
- What are called **soy milk** or **oat milk** are not truly milks, because they are obtained by mixing water

with a crushed vegetable product and the mixture is later filtered. They are, then, ***horchatas*** (Spanish-derived name for milky looking drinks made from ground ingredients). There are also dispersed systems called **suspensions**, made up of water as a continuous medium and small solid particles from the product in suspension, which is what gives them their white color.

The remainder of foods available in the kitchen can be qualified as **objects**, in the sense given earlier to the term. They are whole or chopped biological organisms made up of plant or animal tissues or organs, whole or partial, in turn formed of cells. All preparations that do not exist in nature, and are designed and constructed for various purposes, are also objects, in this case artificial ones. In some cases these are prepared traditionally to take advantage of wastes from slaughtering. In other cases they are prepared to allow food to be preserved over time, like the majority of cold cuts and products from the delicatessen. Other times preparations are made to increase the variety of ways to present the same food: this is the case of the infinite varieties of pasta, all with the same basic ingredients, which are wheat flour molded with a certain amount of water and then dried. And finally, in other cases they are sophisticated preparations for gastronomic pleasure, such as the majority of sweet desserts and candies.

Some of these objects –like seeds for example– can be minced to a powdery state, such as flour, but in its internal structure the powder has the complexity of the organism it derives from. In spite of the apparent similarities between salt, sugar and flour, under a microscope the structures of the starch in the granules can be seen in the latter, as well as their cuticles and the germ if the wheat has not been refined. Conversely, salt and sugar are crystals that, under the microscope, do not show their intimate structure, which can only be deduced with much more powerful instruments.

2.5. Acids, Bases and Salts in the Kitchen

An important chemical characteristic of substances is their **acidity**, and the opposing property, called **basicity** or **alkalinity**. An acid substance and a basic one neutralize each other when they react in the correct proportions, commonly resulting in a neutral mixture. The acidity or basicity of substances is characterized by their **pH** value, which can easily be determined in the laboratory. The pH scale values go from 1 to 14. For historical and chemical reasons, and in contrast to what seems intuitive, value 1 corresponds to the most acidic values, with acidity lessening to 7, which is the neutral value, and the basicity then increasing from 7 to 14, which corresponds to a very alkaline substance.

Alkalizing and Acidifying Diets

The *theory of the alkaline diet*, which has no scientific value, attributes different *acidity or alkalinity* values to foods, apart from their true pH value, according to the supposed effect they will cause in the acidity of blood and urine once ingested. Starting from this false supposition, this theory seeks to deduce if these foods are suitable for the organism or not. Proponents of the alkaline diet, originally based on experiments by **Claude Bernard** in the nineteenth century –correctly done, but poorly interpreted by his followers– connect the acidity or basicity of the foods' ashes, once burned externally, with the predisposition to get certain illnesses. This theory has never been corroborated and, in spite of this, the alkaline diet continues to grow and is defended by many followers of alternative therapies. Suffice to say that blood always has a pH of between 7.35 and 7.45, a value that is nearly unalterable with diet. Conversely, the pH value of urine can indeed vary between 4.6 and 8, but there is no direct relationship between the value of its pH and the appearance of illnesses. Of course diet can influence the pH of urine, but not in a simple way, and as such advice about *alkalizing and acidifying* diets is totally lacking in value. Throughout the following, the terms acid and alkali or base are used in their orthodox scientific meaning.

In the kitchen, practically all plant foods are acids, varying from the most acidic –lime or lemon– to those that are practically neutral, such as peas. Lemon juice has a pH of 2.3, and along with lime juice is the most acidic edible substance that is usually found in the kitchen. Its acidity derives from the *citric acid* it contains. When this acid is used as a food additive it gets called additive 330. Orange juice contains the same acid, but less concentrated, and its pH is around 3.0. All cola soft drinks, including those that do not contain sugar or caffeine, are very acidic (pH 2.4) as they contain a small amount of the additive *phosphoric acid*. Vinegars have acidities ranging between 2.4 and 3.4. Their acidity derives from the content in *acetic acid*. Genuinely natural yogurt contains *lactic acid* as a product from the fermentation of the lactose, the sugar from the milk. Its acidity is between 4 and 5. Sour cabbage, also known as *choucroute* or *sauerkraut*, also contains lactic acid, along with acetic acid and other fermented products. Other common drinks are all more or less acidic: beer has a pH of 4.5; coffee, 5.0; tea, 5.5; and cow's milk, 6.5.

Pure water has a neutral pH of 7.0. Flat mineral waters have pH values that depend on their mineral composition, and can range from 6 to 8. This last value is slightly alkaline owing to the presence of bicarbonate soda and other alkaline salts in the water. The most alkaline substance that is usually found in the kitchen is egg white. In fresh eggs, the pH has a value of around 7.5, but in eggs that are old it can reach values above 9.

The pH value can easily be determined with devices called *pHmeters* or with reactive strips of paper that change color according to the pH value of the liquid. Red cabbage and other edible substances also change their color according to the acidity of the medium. There are chefs who have taken culinary advantage of this characteristic, designing dishes that change color according to the dressing added to them.

Salts are, chemically, substances made up of anions and cations –atoms that have gained or lost one or more electrons forming a crystalline structure that repeats in the entire substance in the form of an undefined tridimensional network. In addition to common salt, in the kitchen we have sodium bicarbonate and different food additives, such as sodium sulfite, an additive used as a preservative in wines and many other foods, or sodium nitrate and potassium nitrate, also preservatives used in canned meats and sausages. Sodium saccharine used as a sweetener is also a salt, as is the flavoring agent sodium glutamate.

As mentioned in section 2.2, salts are found dissolved in mineral waters. As such, the labels for these waters have a list of the anions present, for example bicarbonates, chlorides, sulfates; or cations, sodium, magnesium, potassium, separately. When mineral water evaporates until dry, a residue remains containing the salt crystals present.

Chapter 3
Molecules, the Actors Responsible for Food's Organoleptic Properties

The properties we can detect from foods with our senses, and which are called *organoleptic properties*, are many, with the most important being the following:

- *appearance* and *color*, information provided by our sight
- *aroma*, information provided by the olfactory system
- *flavor or taste*, information provided by taste buds
- *temperature*, provided by temperature sensors in the skin and mucous membranes
- *texture*, information provided basically by touch,

especially in the mouth. Sense of hearing also gives additional information, especially related to the texture –crunch, basically– of the food.

All these form part of the field of **sensory science**, an emerging scientific area.

Knowledge about the chemical composition of a food, that is to say, what molecules it is made up of, can explain its color, aroma and flavor. Conversely, chemical composition is not enough on its own to explain or interpret the texture of a food. Other information is need to do that, especially related with the food's internal structure. All of this will be discussed in the following sections.

3.1. The Color of Foods

We see colors by means of a complex mechanism that involves physics, chemistry, biology and neurosciences. Light from the sun, reflected sunlight or artificial light are electromagnetic radiations having just one color or mixtures of various, with different intensities. The radiation –light– that hits the object we are observing can behave in three distinct ways, depending on the type of radiation and on the composition of the object. The light can pass through the object, and we say that it is *transparent* or *translucent*; it can reflect the light on its surface in a way that is specular –and we say that the object behaves like a *mirror*– or diffusely, if it has a rough surface; or it can absorb all or part of the incident light. Most commonly, an object transmits or reflects a good part of the light received and absorbs another fraction. The part of the radiation that is reflected or transmitted by the object impacts on the retina of the observer. The specialized cells this is made up of transform the light received into chemical signals and then electrical ones that are processed by the brain and interpreted as colors.

According to the chemical composition of the object and its superficial characteristics, different colors can be seen. In 1665, **Newton** observed that sunlight decomposed into seven colors, the same as those in a rainbow, and his classification continues to be used today. It is now known that the visible light spectrum is continuous, that is to say, that we can distinguish infinite colors in white light, not just seven, but in terms of classification the seven basic ones continue to be used, or six, as indigo these days is usually ignored. There are many other colors that are not from the solar spectrum, such as pink or brown. These are colors that are created in the brain circuitry coming from a mixture of signals from various other basic colors. White –which is the sum of all the colors in the solar spectrum– and black, which is the absence of all light, are not from the solar spectrum, either. The mixture of colors from the solar spectrum is seen in the mind as if they were a pure color, having different shades and intensities and, depending on the environment and intensity of the light, one same color can be seen in very different hues.

The majority of colors we attribute to foods are due to the molecules present in them, although there are some exceptions, as we will see. Colored molecules are usually quite large compared with the rest of the molecules in food, and have a chemical structure with links that are sensitive to certain visible light frequencies. These radiations are absorbed by the molecules, while the rest of the unabsorbed frequencies are reflected or transmitted towards the exterior, and are the ones we see as the color of the substance or object. For example, *chlorophyll* absorbs the red and blue tones and reflects the green, and thus is the color seen in plants.

Pigments and colorings are not the same. **Pigments** are powdery solid substances that disperse in a solid or liquid mass without dissolving into it. **Colorings**, conversely, are substances that dissolve in a liquid and dye it a color. Such a distinction is not very relevant in the kitchen, and the terms are frequently confused. Products sold as *food coloring* for home kitchens are usually coloring that is soluble in water or in aqueous mixtures. By contrast, among food additives from the colorings family, there are colorings and pigments. From here on, in this text there will be no distinction made between them.

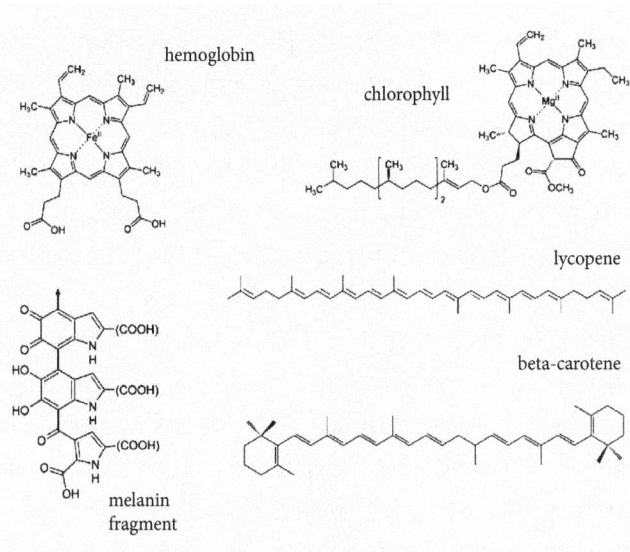

Figure 2. Various food coloring molecules. Chemical structures of the molecules hemoglobin, chlorophyll, lycopene, beta-carotene and a melanin fragment. Notice the similarity of hemoglobin with the right part of the chlorophyll molecule, the first with an iron atom and the second with a magnesium atom in its center. The lycopene and beta-carotene molecules are made up of only carbon and hydrogen atoms, not shown in the figure.

Let's look at different foods classified by color. Many of the natural pigments and coloring they contain are also authorized as food additives, and the INS number of each is given. We will only refer to raw foods, given that cooking can notably change the aspect of a food due to reactions of *caramelization* and the *Maillard reactions*, which will be described in Chapter 8.

There are three large chemical families of molecules that give foods color, especially for plants. *Heme pigments* and *chlorophyll* are similar molecules, with a structure having an atom of iron or magnesium, respectively, in their center. The *carotenoid* family are the most widely-distributed pigments in nature. Those that only have carbon and hydrogen are *carotenes*, and those that also have oxygen are *xantophylls*. *Flavonoids* are another family of pigments, which includes *anthocyanins*, an important compound family related to *tannins*, compounds that give foods astringency and which we will see when speaking about reactions in wines (see section 8.7). In general, each one of these families of pigments can provide extremely varied colors depending on the animal or plant species in which they are produced, and from the pH and temperature of the medium. In Table 3, the main food colorings are shown.

Food additives are pure substances of known chemical composition, and ones that are ingredients in prepared food products. There are different ways of labelling and classifying additives on lists of ingredients. All food additives have a common numbering all over the world, the International Numbering System for Food Additives (INS). In the USA, food additives are usually labelled with their name and must be approved by the U.S. Food & Drug Administration (FDA). In Europe, additives are authorized by the European Food Safety Authority (EFSA) and are labelled with their name and with an E number, made up of the letter E followed by the INS number. In other countries, the same numbers are used without the E prefix. Currently, there are some 370 authorized additives, a number that changes with new incorporations and withdrawals. Prepared foods manufacturers currently try to avoid or minimize the use of additives in their compositions because many users mistrust –unreasonably in the majority of cases– food coloring. As a substitute, what are known as **colored foods** have been developed, which are preparations extracted from foods that have both coloring power and a certain nutritional value. At the time of this writing, legislation has still not defined these substances as food additives, and for this reason they have not been given an E or INS number in the classification. They appear on the list of ingredients with their name: for example, *colored food product (carrot concentrate)*.

Colour	Food	Pigment/coloring	Chemical family	INS Number
Red	tomato	lycopene	carotenoid	160d
	watermelon	lycopene	carotenoid	160d
	red pepper	lycopene, capsanthin	carotenoid, flavonoid	160d, 160c
	red wild berries	anthocyanins	flavonoid	163
	red beets	betains	betains	162
	prickly pears	betains	betains	
	red Swiss chard	betains	betains	
	strawberries	anthocyanins	flavonoid	163
	cherries	anthocyanins	flavonoid	163
	red meat	myoglobin	heme pigment	
	blood	hemoglobin	heme pigment	
	shrimps, lobsters	astaxanthin	carotenoid	161j

Molecules, the Actors Responsible for Food's Organoleptic Properties 61

Orange, Yellow	carrot	beta-carotene	carotenoid	160a
	apricot	beta-carotene	carotenoid	160a
	peach	beta-carotene	carotenoid	160a
	yellow pepper	beta-carotene	carotenoid	160a
	fresh milk	beta-carotene	carotenoid	160a
	butter	beta-carotene	carotenoid	160a
	citrus	beta-carotene	carotenoid	160a
	turmeric	curcumin	polyphenol	100
	wild salmon	astaxanthin	carotenoid	161j
	egg yolk (chicken)	lutein	carotenoid	161b
	corn	beta-carotene and lutein	carotenoid	160a, 161b
	egg yolk (duck)	beta-carotene, canthaxanthin	carotenoid	160a, 161g

Green	green vegetables and fruits	chlorophyll	like heme pigment	140
	raw shrimp and lobsters	carotenoids+protein	carotenoid	
	oil	chlorophyll	like heme pigment	140
Blue, Violet	blackberries and cranberries, purple grapes	anthocyanins	flavonoid	163
	purple asparagus, eggplants	anthocyanins	flavonoid	163
Black	cephalopod ink	melanin	phenolic compounds	
	mushrooms		phenolic compounds	
White	sugar, salt, flour	none		
	milk	none		

Table 3: Food coloring before being processed.

Red-colored Foods

The different colorings from the carotenoids family are responsible for the color of many red fruits. Many are also antioxidants. Some vegetables have more than one coloring, such as the red pepper, which has carotenoids and *capsanthin*, a flavonoid.

The red in meats is due to a heme pigment called *myoglobin* that is in the muscle tissue. The reddest meats –beef– have up to 2% of this compound, while white meats –chicken breast– have less than 0.05%. Tuna meat is also rich in myoglobin. The red from the blood in higher animals contributes little to the color of the meat in animals that are bled. The blood is red because red blood cells have *hemoglobin*, similar to myoglobin. An iron atom is found in the center of both molecules. Molluscs and certain arthropods have blue blood as their pigment is *hemocyanin*, with a copper atom in place of iron.

Cooked shrimp, Norway lobsters, lobsters and crabs have a reddish tone to their shell due to the presence of the carotenoid *astaxanthin* (INS 161j), a natural pigment similar to the *lycopene* in tomatoes, and is also antioxidant. In the live animal, the pigment is associated with proteins from the shell and its color tone is green.

Orange/Yellow-colored Foods

These fruit colors usually derive from *beta-carotene*. This is found in carrots, apricots, yellow peppers and citrus fruits. It is what gives fresh milk its characteristic slightly yellowish tint, a color that remains and intensifies in butter. Beta-carotene is an antioxidant and precursor to vitamin A in the body. In fact, oranges are initially green because of the *chlorophyll* that covers their skin, but changes in temperature and forced maturation in ethylene gas chambers destroy the chlorophyll and orange color tones appear.

Wild salmon has pink meat because it feeds on crustaceans, and as such we can see that their shell contains astaxanthin. In contrast, farmed salmon, which would have grey meat from only eating feed, improve their color when a certain amount of astaxanthin is added to the feed.

The intensely yellow color of egg yolk comes from *lutein*, which the chicken gets from alfalfa and corn. If their diet is changed and corn is eliminated, the yolk from their eggs acquires whiter color tones. Conversely, duck eggs are more orange due to the presence of beta-carotene and *canthaxanthin* (INS 161g), a red pigment they get from aquatic insects and crustaceans.

In general, eggshells are white and are made up of a matrix of protein fibers and calcium carbonate, which is a white mineral. This gives permeability to the gases the egg exchanges with the exterior. When remainders of hemoglobin are also found in the shell, it acquires the reddish-brown color typical of brown eggs. The color depends on the genetic characteristics of the variety of laying hen, but does not affect the nutritional value of the egg at all.

Green-colored Foods

The presence of *chlorophyll* pigment, which has diverse varieties of different shades, in the chloroplasts of green vegetables and seaweeds is well-known. This molecule has a structure that is very similar to hemoglobin but with a central magnesium atom instead of iron.

Blue/Violet-colored Foods

Cranberries and blackberries, as well as some parts of asparagus, owe their blue or purple color to the already mentioned *anthocyanins*.

Black-colored Foods

Black-colored foods are scarce. Squid and other cephalopods use their ink as an element for evasion and escape. This ink contains *melanin*, a black pigment responsible for the dark skin coloring of various human populations and the phenomenon of skin turning dark because of solar radiation. There are also compounds similar to those that give dark-brown coloring to cut vegetables when they oxidize, according to the *enzymatic browning* mechanism, which will be explained in Chapter 8. The dark colorations of many fungi and mushrooms are because of similar compounds.

White-colored Foods

There are some milky white-colored pigments from the families seen previously, but usually white-colored foods are this color because of distinct mechanisms that do not involve the presence of pigments. This is because opaque rough surfaces without coloring, when they receive sunlight or artificial lighting, reflect the light diffusely in all directions. In this case, the object is perceived as white: it is the case of the surface of a white egg. The same thing happens to powdery substances, such as a mass of sugar, flour or salt.

In contrast, there are substances such as milk that are perceived as white-colored by a very different mechanism. Milk is a complex mixture, a dispersed system principally made up of a solution of water with lactose –transparent mixture– that has small drops of milk fat dispersed in the middle –also transparent– and small particles of calcium caseinate, negligible to the naked eye. The incoming light that strikes the milk bounces off the surface of the drops of milk fat, is dispersed in all directions, and when the light leaves the liquid mass it is visualized as white. If the particles are very small, as is the case with homogenized milk, currently the most sold, the dispersion is greater and that is why it looks even whiter. This is the same reason why some beverages are white, both those made from tiger nut (*horchata* in Spanish) and the poorly-named *milks* made of soy, rice, barley, oat or other compositions. In these cases, it has to do with small solid particles from tubers or ground seeds, suspended in the mass of water, with the same whitening effect.

Cooked egg whites look white because the denatured proteins from the egg white encompass the entire aqueous part, resulting in a gel. This gel-like structure disperses light the same way milk does. Denaturing of proteins and gelation will be studied in section 8.1.

The Colors Diet. A good piece of dietary advice is to try to have the most varied diet possible, with foods from all the colors, including white. This is synonymous with a diet having a variety of fruits and vegetables, with their vitamins and antioxidants, and is suitable advice for any type of person. There are, however, some diets based on food colors that make esoteric statements but have no scientific value.

> ### The Fallacy of the Five "White Poisons"
>
> Among people concerned with their nutrition, the five white poisons are often spoken about. These are: milk, refined salt, refined sugar, refined flour and white rice. The arguments to demonize them are varied, but none of them stand up to critical analysis.
>
> Milk is an essential food for young mammals, accepted without discussion. The discrepancy

appears when milk consumption by an adult is advised. The fact that part of humanity is partially or totally lactose intolerant, which is the sugar from milk, and more at advanced ages, has produced an aversion to milk in certain circles. Many milk products are now manufactured without lactose and are consumed by sectors of the population having no lactose intolerance purely due to hypochondria. Some add the argument that human beings are the only mammal that continues drinking milk as an adult. That is true, but it is not a valid argument at all: it is also true that the majority of foods are processed and transformed and human beings are the only ones to eat them, but they are not renounced for this reason.

Also, the calcium contained in milk is better assimilated than what comes from plants. The demonization of milk coexists with the emerging tendency to drink non-pasteurized milk, a practice that can have health risks. Pasteurization or sterilization of milk does not make it lose almost any nutritional value, and increases its safety by reducing or eliminating the population of pathogenic microorganisms it may contain coming naturally from the animal itself or acquired during packaging.

Refined salt is associated with the production of cardiovascular problems, and for this reason is given the attribute of being poisonous. But this also applies to any other unrefined salt, which contains almost the same proportion of sodium chloride. Refined salt does not have all the trace elements, clays and other compounds that some unrefined salts might have, but that does not turn refined salt into a poison and not the others.

Refined sugar is sucrose in a very high proportion, over 99%. Unrefined sugars –molasses, brown sugar, unrefined cane sugar, raw sugar and other varieties– also contain a large proportion of sucrose. If the aversion is due to the presence of sucrose –which produces cavities and has high caloric density– this aversion should also be applied to all the varieties of sugar. The impurities in molasses contain some significant nutrients, but they are in small quantities and as with refined salt can be obtained from the rest of foods in the diet.

White rice is husked. Bran from the cuticle is of nutritional interest as it is an insoluble fiber, necessary in the diet. But that does not make rice from inside the grain –husked rice– poisonous.

Finally, what is said about white rice can be said about refined wheat flour. Depending on the process followed in its preparation, it might not contain the wheat germ or cuticle, and as such perhaps there is not protein –the gluten– or fiber in its nutritional composition. Other flours are excellent from the nutritional and gastronomic point of view, but that does not make it necessary to demonize white flour. An additional health aspect is the incidence of celiac disease, where those affected cannot consume wheat or its derivatives. But this includes white

> flours as well as wholemeal ones, and also ancient and less frequent varieties of flour such as spelt or kamut. Celiac disease or gluten intolerance is an autoimmune disease of genetic origins that requires completely avoiding gluten. But, like with milk, many consumers of products for celiacs are not really celiacs, and they limit their diet unnecessarily due to hypochondria or thinking that in this way the gluten will not trigger the disease in them at some later date, a useless precaution from a clinical perspective.

3.2. Smell, Taste and Flavor in Foods

In human beings, the sense of smell and sense of taste are anatomically differentiated. The *pituitary* on the one hand and *taste buds* on the other produce independent signals. But the oral cavity and nasal cavity communicate by means of the pharynx, and the molecules that the sense of smell is sensitive to arrive by direct inhalation through the nasal passages –via *orthonasal*– as well as by the via *retronasal*. There are molecules that interact with both olfactory and gustatory receptors, and there are substances that have different molecules, some of which activate the sense of smell and others the sense of taste, providing a complex sensation. For these reasons, the concept of *flavor* adds actually the characteristics of aroma and taste, and flavor is a feeling that integrates both.

The term *aroma* may create confusion. Daily language refers to the smell of a substance, but for the food industry an aroma is any substance that modifies the flavor or smell of a prepared food. In accordance with most legislation, it is not necessary to label which aromas have been added as ingredients to a processed food: it is sufficient to mention if the food contains them, with the understanding that they form part of the list of authorized aromas in the food authority's registry. Certain aromas do indeed need to be explained: for example, in cola drinks it is necessary to indicate the presence of *caffeine*, which in addition to adding a bitter taste is a stimulant; in tonic water, it must be specified if it contains *quinine*, also a bitter substance that can cause undesired cardiovascular effects.

There were four classic flavors it was considered that the palate and tongue could detect: sweet, salty, sour and bitter. The substances that were typically associated with each one of these flavors were sugar, table salt, vinegar and some drinks such as coffee without sugar. It was thought that each one of these flavors was noted primarily by an area of the tongue, but today this idea is considered only partially true. In recent years various other flavors have been added to these "historic" ones: the flavors of *umami*, spicy, astringent and fat. All of these derive from countless molecules or chemical substances present in foods, of which a few will be cited.

Sweet Substances

Sweet substances usually contain one of the natural sugars produced by plants in photosynthesis. The principal simple sugars or **monosaccharides** are *glucose* and *fructose*, in various fruits, and *galactose* in milk. **Disaccharides** have this name because they can be considered a union of two monosaccharide molecules. Among these are *sucrose* or table sugar, the *lactose* from milk and *maltose* from beer. Additive sweeteners from the 950 series and those following (Table 4) also give a sweet flavor. Molecules known as **polysaccharides** also exist, formed by the union of various or many monosaccharide molecules. Among them are *starch* and *cellulose* from plants, which do not have a sweet taste and are not directly assimilable by humans in the digestive process; and *glycogen*, a reserve substance made up of many glucose units and stored in the liver and other organs in animals.

> ### Sweeteners
>
> A sweetener is a substance that gives a sweet taste to other. Natural sweeteners are usually sugars, from the carbohydrates family, and have a caloric value of 4 kcal/g. The principal and most widespread is sucrose, which is beet or cane sugar. The sweetening power of a product is compared to that of sucrose, which is given a value of 1 on the sweetening scale.

> With other sweeteners, their sweetening power is given in parentheses. This value has a certain degree of uncertainty in many cases.
>
> Other caloric sweeteners derived from natural products are, among many others, honey (0.95), glucose (0.7), maple syrup (2), fructose (1.8) or invert sugar (1.3; see Chapter 11). Corn syrup with a high fructose content is obtained industrially from starch or cornstarch. This has substituted sucrose in many sweet soft drinks.
>
> A large group of non-caloric sweeteners have been developed, with the goal of fighting obesity, reducing cavities, and making the ingestion of sweet foods possible for people suffering from diabetes.

150a Plain Caramel. Caloric, classified as brown coloring
150d Sulfite Ammonia Caramel. Caloric, classified as brown coloring
420 Sorbitol and Sorbitol Syrup (0.6). Caloric, derived from sugars
421 Mannitol (0.7). Caloric, derived from sugars
950 Acesulfame K (130-200). Non-caloric, synthetic
951 Aspartame (100-200). Non-caloric, synthetic
952(ii) Sodium cyclamate (30-80). Non-caloric, synthetic

953	Isomalt (0.4-0.6). Caloric, derived from sugars
954(ii)	Saccharin (and Na, K, Ca salts) (200-700). Non-caloric, synthetic
955	Sucralose (600-800). Non-caloric, synthetic
957	Thaumatin (2500-3500). Non-caloric, natural
959	Neohesperidine DC (2000). Non-caloric, synthetic
960	Steviol glycoside (300-400). Natural, non-caloric. Is commonly known as *stevia*
961	Neotame (7000-13000). Non-caloric, synthetic
965	Maltitol and Maltitol Syrup (0.7). Caloric, derived from sugars
966	Lactitol (0.4). Caloric, derived from sugars
967	Xylitol (1). Caloric, natural or derived from sugars
968	Erythritol (0.7). Caloric, derived from sugars
969	Advantame (37000). Non-caloric, synthetic

Table 4: Principal sweeteners classified as food additives by order of INS number.

All synthetic sweeteners are approved by food safety agencies, although some are not authorized in certain countries for historical or commercial reasons.

Salty Substances

Table salt, chemically sodium chloride with some impurities, is the substance that gives the salty taste to all salted foods, in which it also acts as a preservative, and is a universal ingredient in many stews and prepared dishes. The diverse brands and types of salt have somewhat distinct tastes due to the presence of other salts that crystalize along with the sodium chloride. The most habitual is *magnesium chloride*, which gives it a certain bitter flavor. Other salt, such as *potassium chloride*, is used as a substitute for common salt or mixed with it in salts prepared for people with high blood pressure. They have a salty taste but are clearly distinguishable from the normal salty taste.

Sour Substances

The *sour* flavor –nearly synonymous with *acidic*– is typical of foods that have gone through acetic fermentation, such as vinegar, as they contain *acetic acid*, a derivative of ethyl alcohol (see section 8.2). There are other foods with acidic molecules in their composition: citrus fruits have *citric acid* (INS 330) and apples, *malic acid* (INS 296). Yogurt and other dairy products, in which lactose from the milk has become *lactic acid*, and *choucroute* or *sauerkraut*, whose sugars have also fermented into lactic acid, are also sour substances. The drink known as *kombucha*, popular in U.S. society for a number of years now, is the product of fermentation of sugared tea that ends up creating a somewhat sugary sour mixture, with a bit of ethyl alcohol and acetic acid.

Bitter Substances

Compounds present in foods having a bitter taste are extremely varied. Coffee and cocoa beans contain *caffeine* and cocoa also has *theobromine*. Both have a characteristically bitter flavor. Other foods that are bitter at the source are endive, bitter almonds –used in the preparation of sweets or liqueurs– as well as grapefruit, or olives before being cured. Various prepared drinks are also bitter: beer, tonic waters or bitters. The bitter taste of beer is due to formation during the cooking process of certain acids derived from compounds present in the hops. Bitters were first sold as medicinal potions but are now simple aperitifs, with or without alcohol. There are digestive liqueurs whose bitterness comes from the compounds present in the plants they are made of: angostura, gentian or bitter orange, among others. *Quinine*, a natural aroma extracted from the bark of the cinchona tree, is an ingredient of the tonic water and is also used as a malaria medication. Quinine is chemically an *alkaloid*. These are organic compounds from various chemical families having nitrogen in their molecule. Almost all of them have plant sources. They have a metabolic action, very different depending on the substance, and are used as medicines or drugs, legal or illegal: *atropine, codeine, ephedrine, strychnine, LSD, morphine, cocaine, mescaline, quinine, caffeine, nicotine* and *theobromine* are all alkaloids.

Umami-flavored Substances

Umami is a Japanese term that means *tasty*. This flavor was discovered in 1908 but was not globally accepted as a universal flavor until the 1980s. It is the characteristic flavor from *monosodium glutamate*, a substance present in various foods such as soy sauce, parmesan cheese, mushrooms and other fungi, or ham. It is sold as a food additive and classified as a flavor enhancer. Its taste increases due to the presence of *inosinate* from sodium or potassium, substances present in foods such as meats which are also sold as additives 631 and 632.

Fat-flavored Substances

Fat flavor or *oleogustus*, as it is sometimes called, is the sixth flavor, accepted by experts in 2015. It is typical of very greasy substances such as bacon or smoked bacon, and is attributed to the presence of fatty acids in the food. It appears there are specific sensory receptors for this type of molecules. Not all scientists consider this flavor to be well-defined for the moment.

Spicy Substances

Spicy is not considered a flavor but rather a painful feeling of irritation of the mucous membranes, which is assimilated as a flavor since the mouth is its entryway and because it attacks the same areas where flavors are detected. But it is not detected by taste buds but rather by other sensory receptors called *nociceptors* or pain receptors. Spicy is without a doubt one of the basic feelings in universal cuisine. By via retronasal it also reaches the pituitary, where it produces strong, at times very painful, feelings. The feeling of spicy is called *pungency* or *spiciness*, and measurement scales have been developed that compare the spiciness of diverse substances, with the *Scoville scale* for measuring the pungency of different varieties of peppers being the most well-known. This scale begins at zero for sweet peppers and reaches 15,000,000 for pure *capsaicin*, which is the spicy molecule from different varieties of peppers. There are inedible substances that are even spicier, such as *resiniferatoxin*, one thousand times spicier than capsaicin. Black pepper, certain spicy peppers, chili peppers and mustard are common in Western kitchens. Common in other kitchens are *jalapenos* (2,500 on the Scoville scale, like spicy peppers), *chipotle* chili (10,000) or *habanero* chili (a minimum of 350,000). Another well-known spicy food is German horseradish, called *wasabi* in Japanese. Ginger is a rhizome that increases its spicy flavor as it dries. Its spicy molecule is *gingerol*, with a molecular structure similar to capsaicin.

Astringent Substances

There is no agreement among experts to consider *astringency* a flavor, as it also has components of a texture. It is the complex residual feeling that remains in the mucous membranes and tongue after having certain substances such as strong tea, drinking certain red wines or biting the peel of a banana or the skin of a green kiwi. The majority of astringent substances contain substances called *tannins*, which are compounds that act as plant defenses.

There are other taste sensations, such as the taste of *freshness* of substances even when they are not cold due to compounds such as *menthol*. It is thought they activate the trigeminal nerve, but without any specific taste receptors. The feeling of *metallic taste*, like what some artificial sweeteners have, is connected with the electrical sensation felt when biting a metal and which has its origin in the creation of electrical microcells in the humid environment inside the mouth.

Flavor is the sensory impression resulting from the integration of taste, aroma and, partially, the feeling a food has in the mouth. *Flavorists* are professionals with knowledge on how to modify the flavor, aroma and texture of a food to get an improved product in respect to the original (Segnit, 2010; Shepherd, 2012).

Novel Foods

Diverse edible substances that had not been used significantly as foods by the population until a certain point (in the EU, before 1997) are classified under the name of novel foods. This includes foods imported from unusual and third countries, foods and ingredients having a novel molecular structure, and modifications of products not considered food before and which have been modified to make them edible. Some of these new foods are insects (see Chapter 11), genetically modified products (which have specific legislation) or exotic fruits and vegetables, traditional in their country of origin. Foods that have been processed with new preservation techniques –such as electrical pulses, magnetic fields and others– are also considered novel foods.

Products hoped to be marketed and which have been classified as a novel food must be subject to a strict authorization protocol, which usually lasts various months or years. Third countries consider this to be a disguised way of establishing trade barriers. These products are subject to evaluation by health authorities for their eventual authorization for consumption.

Some of these novel foods are packaged products with brand names, such as certain yogurts that have phytosterols and stanols in their composition.

These are molecules derived from plants that compete with cholesterol in intestinal absorption so that blood cholesterol levels are reduced. Other plant derivatives are fruit pulp from the baobab, or fruit from the maqui, a Chilean and Argentinian phanerogam considered a good natural antioxidant. Stevia, a plant with a powerful sweetener on its leaves, was considered a novel food until the extract from its leaves, which contains steviol glycosides, was approved as a sweetener (INS 960). In Europe and the U.S., for the moment, marketing of the plant for infusions or as a dietary supplement is not authorized as studies on its safety have not been finished.

Designer foods are foods obtained by genetic engineering techniques, or improved with special additives. The process to obtain them is known as fortification or nutrification. Advertising for some of these types of foods calls them *superfoods*.

Chapter 4
Biological Structures. Where are the Nutrients?

Living beings are not the same as plastic bags, where the chemical compounds they are made of are all mixed together. On the contrary, there is a strict structured organization. Let's try to understand where the chemical compounds are placed in the cells. In general terms, we can imagine a cell from any living being as formed by a container that is its *membrane*, constituted by a set of structured molecules as an impermeable wall with pores. Inside the membrane there are different organelles such as the *nucleus*, the *mitochondria* or, in plants, *chloroplasts*, among others. Each one of these organelles is made up of diverse molecules, and all of them are submerged in a liquid called *cytoplasm*. This liquid is basically water that contains other molecules that are dissolved or suspended. This is of interest in the kitchen because many cooking processes consist of trying to soften cell membranes or in coagulating the liquids inside of them, and knowledge about their structure helps to understand the phenomena that take place in cooking. But not all cells are equal, and their structure depends on which animal or vegetable species they are, and which tissue or organ they form a part of.

Green plants or phanerogams are made up of, in general terms, roots, stems and leaves, which are support for flowers and fruits with their seeds. The generic plant cell has a membrane made of *cellulose* and *proteins*. Inside they have some ample spaces called *vacuoles*, full of water having dissolved salts and sugars. Over time, the vacuoles of a cell join together and increase in size. *Chloroplasts* are inside the cellular cytoplasm, and inside of them is where the photosyntesis is carried out. That is why they contain *chlorophyll*, the pigment that gives them their green color, present above all in the leaf cells. Root cells do not have chloroplasts, and many species (carrots, turnips, parsnips, radishes) can store *starch*, a carbohydrate similar to sugars but with much larger molecules (see Chapter 8). The cells from some specialized stems can also store starch and sugars: tubers (potato or sweet potato), rhizomes (ginger) and certain bulbs (onion, garlic).

Fleshy fruits are usually distinguished from dried fruits. In the first ones, such as the peach or pear, the cells of its external membrane or *pericarp* accumulate sugars and water, and also oils and fats like in the case of olives and avocados. What are called **dried fruits** usually applies to nuts/seeds as well as dried fruits. As such, raisins and dried figs are dried fleshy fruits that, having lost water, have concentrated their sugars. Conversely, in hazelnuts or walnuts the pericarp cells are dead and envelop the seeds, which accumulate proteins and fats for development of the new plant. These are the typical dried fruits, although there are ones having different morphologies, such as sunflowers or pumpkin seeds.

Cryptogram plants do not have chlorophyll. On the cell walls of many of their species, such as fungi, there is *chitin*, which is a carbohydrate also present in the exoskeleton of crustaceans and insects. It is similar to cellulose and gives rigidity to the structures it forms part of.

The **animal cell** is very different, in general terms, from the plant cell. It does not have vacuoles nor chloroplasts, and is very specialized depending on the tissue or organ of which it forms part. As such, in the **adipose tissue** the cells or **adipocytes** accumulate fat –*triglycerides*– in their cytoplasm. Adipose tissue is found in the deepest layer of the skin, covering some organs, or in bone marrow. **Muscle tissue** makes up a good part of the mass of proteins in an animal and is formed of long cells that, for this reason, are usually called muscle fibers. These can contract, some muscles voluntarily and others automatically, which permits the movements and displacements of animals. They are basically formed of extremely varied proteins.

Skin and **epitheliums** cover all organs and spaces in the body. Skin is the largest organ in the human body –some 5 kg in an average adult– and is present in all higher animals. Its cells are arranged in a layered structure. The most external has a resistant protein called *keratin*, which is also in all the skin's external structures, such as nails, horns, hair and feathers. In a certain way it has a structural function, like chitin. *Collagen*, a protein that makes up 25% of the weight of proteins in mammals, is found in internal layers of skin like connective tissue. Its degradation with temperature, giving *gelatin*, is one of the keys to cooking meats, as will be seen in Chapter 9.

Finally, **blood** is a type of liquid connective tissue, formed by an aqueous colloidal suspension, with red blood cells, white blood cells or leukocytes and platelets suspended in the blood plasma, which is an aqueous solution with glucose and many other components. We should remember that blood's red color is due to the *hemoglobin* pigment from red blood cells, and that this is very similar to chlorophyll but with iron instead of magnesium.

Supply of Nutrients

The **nutrients** we take advantage of from living beings are chemical substances distributed in specialized parts of plants and animals. As in humans, proteins and fats are found principally in the muscles and adipose tissue. But in contrast to plants, we animals do not have anywhere to store large quantities of carbohydrates. Only the liver can accumulate significant quantities of *glycogen*, a carbohydrate derived from glucose sugar. This is why we have to resort to fruits, seeds, stems and roots from plants for our alimentation. We also find fats in plants and in certain fruits and seeds, and proteins, which make up different tissues. Other nutrients such as vitamins, minerals and fiber are found mostly everywhere, and a varied diet satisfactorily provides us with all the required nutrients.

Chapter 5
Food texture

It is probably **texture**, perhaps more than flavor, the most highly sought property in prepared foods, at restaurants as well as in home kitchens or the food industries. We strive for a steak to be tender, that cream is…creamy, that ice cream does not have ice crystals, that cake is spongy, that a pear is not too grainy. All these terms refer to texture.

It is very difficult to summarize in a few words the concept of food texture. Texture is the characteristic a food has that is perceived by the senses, and especially by touch, during its ingestion and swallowing. There are dozens of terms to characterize textures, and not all authors coincide in their nomenclature. Terms such as brittle, fragile, elastic, rubbery, thick, viscous, moist, powdery and many many more refer to aspects of food texture. One single food might have various textures simultaneously.

Picture 2. Chocolate. Various textures of the same foodstuff.

There are three large groups of properties that help define food textures, especially of solids: **mechanical** properties, **geometric** properties and **surface** properties. The standards of different countries set diverse qualifications that characterize the texture of a food. For example, among the mechanical properties is hardness, with descriptors such as soft, firm and hard, and examples for each one: cheese spread, olives and hard candy. Or elasticity, with descriptors like plastic (lard) and elastic (squid). Some geometric properties are the graininess (mealy: powdered sugar; grainy: semolina; gritty: some pears; lumpy: soft white cheese; pearly: caviar) or the structure. And among surface properties, the moisture or a fatty nature. As such, an apple would be firm, crunchy, chewable and moist, while a canned sardine would be chewable, plastic and oily. This same classification exercise can be done for each food.

All the properties that define an aspect of food texture can be determined by means of some more or less complex experimental techniques. **Texturometers** are devices that try to simulate the force applied on foods during mastication. The forces necessary to deform, cut or grind the food and the variation of force required over time can be measured with these devices. All this allows quantitative characterization to be made of some of the characteristics of their texture. Determination of textures can also be achieved by means of **tastings** done by panels of experts, and their results can be compared with those obtained by the laboratory techniques mentioned.

Foods consumed without the need for large transformations, such as fruits, have textures corresponding to their nature. But the rest of cooked foods have textures achieved by means of the culinary procedures applied to the ingredients. Grinding and mixing of components allows new substances and objects to be obtained that were previously non-existent. In cooking processes the composition of food frequently changes, by which the molecules present are modified and as such flavors, aromas, colors and also food textures change. The changes in temperature usually associated with these processes, as well as the mixture of products, can notably change their textures as well: as will be seen, when a food is heated its fats melt, collagen reduces to gelatin, gases and vapors can appear, and flours hydrolyze. In a certain way, it could be said that one of the principal objectives of gastronomy and of cooking is to obtain the desired compositions and, above all, textures from some given ingredients.

Textures derive from the inner structure of foods, which includes their chemical composition and microscopic structure. The immense variety of foods can be reduced to four large typologies, which will be succinctly described and remarked upon. These are liquids, solids, powdery materials and dispersed systems.

5.1. Liquids: Solutions

As seen in the previous section, from the chemical point of view the majority of liquids present in the kitchen are pure or almost pure substances such as water, or simple solutions: wine, oils or vinegar. The properties of liquids are well-known: they flow with greater or lesser ease and maintain their volume but not their form. They are characterized by their *density* and *viscosity*. This second property is the resistance a liquid shows to flowing, and from the culinary and gastronomic point of view is much more important than its density. The less viscous liquids in the kitchen are water and concentrated alcoholic solutions, such as liquors without sugar (vodka, whisky, gin and distillates in general). Fruit juices –which contain some pulp and natural as well as added sugars– have greater viscosity. Finally, the most viscous liquids in the kitchen are oils. Even more viscous substances, such as gelatins, require a separate classification and are included in the dispersed systems.

Viscosity is determined by the size of the molecules and the interactions between them: water molecules are the smallest of all those cited, but due to their molecular structure each molecule is weakly joined to those adjacent, which gives liquid water greater viscosity than what it would otherwise have due to the size of its molecule. The molecules of any oil are dozens of times larger than those of water, and although interactions between its molecules are much weaker than those of water, their greater size and branched shape cause oils to have much greater viscosity.

The viscosity of any liquid reduces greatly as temperature increases due to the higher mobility of the molecules in hot liquid. This becomes apparent when heating an oil, which is not very fluid cold but is much more fluid at frying temperature, usually above 300°F. Inversely, when cooling a sugarless alcoholic drink such as vodka in the freezer, an increase in its viscosity can perceptibly be observed.

Not all liquids are solutions. Milk, for example, is an emulsion that is a dispersed system that will be described later (see Chapter 6). Other more or less liquid substances such as liquid caramel or certain sauces will also be seen as dispersed systems.

5.2. Solids

The large majority of adjectives that describe textures refer to solid substances. From the mechanical point of view, the elasticity of food has special importance, because when chewing any food the hope is that it can be crushed with the teeth and molars, and that it will not go back to its form when you stop putting pressure on it, as if it were a rubber band. There are three basic behaviors of substances when facing forces of crushing, stretching or cutting: *plastic* substances such as butter reduce and do not recover their original form; *elastic* substances such as meats or squid meat stretch, bend and contract, and return almost to their original shape when the force stops; and *nondeformable* substances, such as seeds or toast, break. Many foods show intermediate behaviors among the three.

At the end of Chapter 4 we confirmed that the biological structures among them are very different. Three examples will make this enormous variety apparent. The **meat** from mammals, fish and fowl, whose basic function is to move the organism or its parts to go from place to place or do some sort of action, consist of groups of elongated muscle fibers linked by protein materials such as collagen, more or less wrapped in fatty tissues. Each fiber is formed of millions of adjoining structured cells, and they can contract or stretch under action from external stimuli. Their behavior comes from their biological structure, not so much from their chemical composition. In contrast, plant tissues from **vegetables** can be hard or soft depending on their position in the plant and their age, and are usually flexible. **Seeds**, cereals and other grains, conversely, act as a reservoir for complex carbohydrates and fats for the future plant's development, and for this reason their cell structure consists of cells full of vacuoles where starch is found in crystal shapes. This structure contains little water and as such the seeds and grains are usually hard, relatively fragile and brittle.

Solid or semi-solid prepared foods can have much more varied textures. There are a pair of forms that deserve special attention. These are powdery materials and the dispersed systems. Because of their importance, the latter will have its own chapter.

Powdery materials are those that present such a high degree of disaggregation that each of their particles cannot be individualized from the others. **Crushing** is to break an object and reduce the size of the fragments obtained to a relatively large grain size, on the order of 0.4 inches. **Grinding** is to reduce the size of the grains to values of 0.04 inches (a grainy or gritty texture) or less than 0.0004 inches (powder or impalpable powder). In any case, there is no well-defined limit of when a material can be considered a powder or not: there are intermediate cases, such as finely grated crackers or sugar. In the latter, sugar in crystalline form can be pulverized more to become *icing sugar*, also called *powdered sugar* or *confectioners' sugar*, whose grains have a dimension of less than 0.004 inches.

Practically all powdery products in the kitchen derive from crushed solids. As has been seen, salt or sugar come from highly pure, fragile and brittle crystalline substances, ground to the desired grain size. All flours are the product of grinding seeds or grains. The behavior of powdery masses is very different from that of the foods they come from. Their texture is defined by greater or lesser graininess and by adherence to the palate or stickiness. *Graininess* is defined by the range of sizes of their particles, which can fluctuate between wide margins, and depends on the utensil used for grinding and duration of the operation. The *adherence* of a powdery substance to the palate derives from its chemical composition: flour or cocoa powder in contact with the surface of the mucous membranes in the mouth leave a big feeling of stickiness and dryness. In contrast, powdered sugar is very soluble in saliva and gives instead a feeling of freshness. The cocoa paste and sugar used in the preparation of chocolate must be very finely ground to ensure that the texture of the product in the mouth, when it is melted, is absolutely smooth and not grainy.

5.3. Texturizers

All elements have their own texture, but on many occasions cooks prefer to change it: for example, if a thicker sauce is desired, potato or corn *starch* is added. Substances that change the texture of food are called **texturizers**. Although they are as old as cooking, in recent years they have become notably more popular. *Gelatin*, an animal protein obtained from fishtails or from the connective tissue of pigs or cows, has been used since ancient times to get jellied textures. Starches from diverse sources are thickeners; *pectin* is used in marmalades. Pectin and starches are both polysaccharides, that is to say, long chain carbohydrates obtained from fruit skins and from tubers or stems, respectively.

> **Texturizers**
>
> As of 1998, experimental chefs have used a wide range of texturizing products, mainly obtained from land plants or algaes, among other reasons to encourage consumption of their preparations by vegetarians and Muslims. Some examples of texturizers are agar-agar, used as a food additive (INS 406) and which is a polysaccharide obtained from algaes; *gellan gum* (INS 418) is another polysaccharide obtained from a bacterial culture; *methyl cellulose* (INS 461) is obtained from

> cellulose; and having huge culinary importance, *xanthan gum* (INS 415), whose value lies in the fact that it is capable of thickening and gelation in cold. *Alginates* (INS 400 to 405) are the key to gelations known as spherifications, which will be described in section 8.5. All these products have applications not only in highly specialized gastronomy, but also in designing diets for people with problems such as dysphagia –difficulty in swallowing– or having certain intolerances.

Xanthan gum, one of the most important in current modern cooking, is obtained by the fermentation of sugars with the *Xanthomonas campestris* bacteria. It shows thickening properties with cold and heat, and gives great viscosity to the liquids it is added to, which can be acidic, alkaline or saline. This has allowed the design of two novel culinary applications. A liquid with xanthan shows a *suspensory effect*, and can retain solid or semi-liquid particles without them settling, a property that is valued in certain dishes or in cosmetic products. It can also retain gas bubbles, and for this reason some producers prepare sparkling wine with xanthan, apparently a liquid but with bubbles on the inside that do not rise easily.

Chapter 6
Dispersed Systems, the Key to Many Culinary Preparations

It has always been repeated that there are three states of matter: **solid**, **liquid** and **gas**. The process to go from one state to another, by supplying heat, are the following: **fusion** (solid to liquid) and **vaporization** or **boiling** (liquid to gas or vapor). It is also possible to go directly from solid to gas, by means of **sublimation**. The inverse processes are, respectively, **freezing**, **condensing** and **desublimation**. The difference between vaporization and boiling is that the first is the creation of vapor at whatever temperature, more so the hotter it is. In contrast, boiling is the brusque and turbulent vaporization of a liquid, where bubbles appear throughout its mass, at a characteristic temperature for each pure liquid and for each atmospheric pressure. Thus, at sea level water boils at 212ºF, or liquid water can exist at atmospheric pressure and at sea level above this temperature. Pure ethyl alcohol boils at 172ºF, but can be smelled when cold because it vaporizes at any lesser temperature. An oil would have a very high boiling point, above 480ºF at atmospheric pressure, but before reaching this temperature different chemical phenomena would have taken place that make the oil decompose, as will be explained further on. Sublimation, or water going from solid to vapor, takes place only at temperatures under 32ºF and is normal in cold areas on the planet. Some modern techniques to dry foods are based on this process, such as **lyophilization** (freeze-drying), which will be described in Chapter 10.

The solid-liquid-gas classification is valid only for some pure substances such as water but not for biological materials, and among these are food products. In fact, the reader should look at how many of the substances around them melt when heated. Paper or wood or plastics, or any food, won't melt but rather will burn first. And the few that melt, such as wax or gelatin, do not boil but instead decompose first. This is because of their composition and the microscopic structure of their components. To describe the properties and the changes of state of the group of substances present in foods and living beings, the concept of **soft matter** has been coined, as differentiation from the hard matter represented by metals, ceramics and glass. Some of these soft materials are normal substances in foods, such as emulsions, foams, suspensions and gels.

From the chemical point of view, all these structures receive the name of **dispersed systems**, **colloidal systems** or **colloids**, as their properties are in some cases analogous to those in some natural glues ("kolla" in Greek). A **dispersed system** is an intimate mixture of two or more substances having a different chemical nature –oil and water, for example– where one cannot be dissolved in the other, but when given enough mechanical energy, by shaking or other means, can be intimately mixed, with the mixture maintaining a certain stability for a time without becoming a solution. The dispersant presents continuity throughout its mass, and for this reason is called the **continuous** or **dispersant medium**. The dispersed substance is in the form of drops, granules or bubbles and is called the **dispersed phase**. Although the contrary could be assumed, the continuous medium is not always in greater quantity than the dispersed phase. The continuous medium can be water or a watery substance and the dispersed phase oil, or the reverse: both cases are possible and common in practice.

In a strict sense, all dispersed systems are unstable and after more or less time are destroyed, with the mixture's components separating and settling in the container by order of densities. It is desirable to increase the duration of these dispersed systems, which can be achieved by different processes. A stabilization process consists in trying to make the particles, drops or bubbles of the dispersed phase as small as possible. As such, the area of contact between the phases does indeed increase a lot and separation by densities is more difficult: a microdrop of oil should rise in the middle of the aqueous phase, but the smaller the drop is, the greater the friction opposing it. This is why milk in containers is stabilized by means of a process called **homogenization**. This process consists in making the milk circulate by means of microholes in a metallic block, by which the drops of fat that are larger than the diameter of the holes in the block break up as they pass through it and are divided into smaller drops, which makes the final emulsion more stable. A secondary consequence of this process is that the milk is appreciably whiter than before being homogenized, which was mentioned before when talking about food colors.

Alternatively, stabilization of emulsions and other dispersed systems can be achieved by means of using additives known as **surfactants**. These are substances whose molecule has a part that is allied with water and the opposite end allied with oil. Their presence in the emulsion acts as a stabilizer because it submerges one end of the molecule in the drop of oil and the other end in the water mass, acting as an anchor between both phases. There are many types of stabilizing additives, which are called *emulsifiers*, some of which appear in Table 5.

322	Lecithins
401	Sodium Alginate
406	Agar
407	Carrageenan
410	Carob Bean Gum
412	Guar Gum
413	Tragacanth Gum
414	Arabic Gum
415	Xanthan Gum
418	Gellan Gum
425	Konjac Flour
432	Polyoxyethylene (20) Sorbitan Monolaurate
440	Pectins
460	Cellulose
461	Methyl Cellulose
466	Sodium Carboxymethyl Cellulose
470a	Salts of Sodium, Potassium and Calcium from Fatty Acids
471	Mono- and Di-Glycerides of Fatty Acids

Table 5: Some emulsifiers, thickeners and gelling agents, classified by order of their INS number.

Foam is a dispersion of a gas in a liquid, in which this liquid is in the form of fine membranes enveloping the gas, which is like bubbles that are independent from each other. Some culinary examples are whipped cream, beaten egg whites and meringues. In current haute cuisine, foam obtained from manual whipping is usually called **air**, while the term **foam** is usually reserved for the gas-liquid dispersed system obtained by using a device in the kitchen called a **siphon**, a closed recipient in which pressurized gas (normally carbon dioxide) disperses in the middle of a liquid or dough, and the mixture is pushed out of the utensil by gas pressure in the form of foam. The density of foams and airs is very low.

In the industry and on the household level, foams are prepared with devices called **aerosol sprays**, analogous to the siphons described. The term **aerosol** means the dispersed system of microdrops from a liquid in the air, like what is produced by nebulizers and sprays for perfumes or medication. Aerosol sprays that dispense whipped cream, cheese spread and other food products can be found on the market. Foamy textures do not appear on most texture lists as they are considered semi-liquids. Some food foams, such as whipped cream, are stabilized by means of additives, like those derived from fatty acids. A high concentration of sugar in the liquid also stabilizes the foam as it increases the viscosity of the liquid, and the presence of certain proteins from the mass likewise act as stabilizers given that they locate in the gas-liquid interface, increasing the resistance of the membranes.

Solid foams are dispersions of bubbles from a gas in the middle of a solid. Examples of these are cooked dough in breads or sponge cakes. They usually have an elastic texture, especially if the flour they are made from has sufficient *gluten*, a protein that gives dough elasticity. Gas is usually produced by a chemical reaction inside the dough when baked, whether due to the presence of biological yeast, or by means of chemical leavening that decomposes with temperature.

An **emulsion**, like those previously cited, is a dispersed system in which a liquid is dispersed in the form of drops in the middle of another liquid that is immiscible with the first. The examples of milk and butter, emulsions of oil in water (O/W) and water in oil (W/O), respectively, were seen earlier. Although the two phases of an emulsion are usually liquid, at low temperatures their texture can become thick or almost solid, like in butter or margarine. The textures of emulsions are characterized by their viscosity, which in qualitative terms can be *fluid* like water, *thick* like a cup of hot chocolate, or *viscous* like condensed milk or honey. Many sauces are emulsions and are usually stabilized with surfactants, at times naturally present among the ingredients. Thus, *mayonnaise* is an emulsion made of drops of oil and fat from the yolk dispersed in the middle of the water from the egg white. Stabilization comes from the surfactant *lecithin*, a fat present in egg yolk and soy, which is sold as a food additive (INS 322). Lecithin molecules coat the drops of oil, isolating them from the water in the egg white and the yolk, which is the continuous medium. The two grams of lecithin in an average yolk can –theoretically– help prepare up to three liters of mayonnaise, although in daily practice that is not feasible. It is somewhat more difficult to prepare **aioli**, as in its original form it is only made of garlic, oil and salt, but commonly some egg yolk is added. In this case, the emulsion is based on drops of oil wrapped in water coming from the crushed garlic, which contains *alliin* that becomes *allicin* when put in contact with oxygen from the air and some enzymes contained in the garlic itself. Allicin and alliin are sulfur compounds with certain surfactant and emulsifier properties. **Lactonese** is an emulsion like mayonnaise, but somewhat more fluid, in which milk substitutes for the egg.

A **suspension** is a dispersed system made up of solid particles in the middle of a liquid. An example is *bechamel sauce*, where flour particles are cooked in milk and butter, changing their chemical composition and, above all, the texture of the mixture, which becomes semi-solid. *Hot chocolate* is also a suspension in which microscopic solid particles from the cocoa paste –obtained from grinding the cocoa bean– and from finely ground sugar are dispersed in an emulsion formed by the milk and cocoa butter. Sometimes an emulsifier like the previously-cited lecithin can be added. This operation of mixing the ingredients from chocolate is called **conching**.

Chocolate in bars is a **solid dispersion** in which very fine solid particles from cocoa paste and sugar are dispersed in the middle of cocoa butter, a fat that is partially melted at room temperature. As a whole, it is a solid that is easily melted, fragile and brittle. *Cocoa butter* is a complex mixture of triglycerides that crystalize in six different forms at around 93°F, which would give chocolate different textures. To ensure its uniformity it is **tempered**, which consists of melting the chocolate, cooling it to 82°F and then heating it again to approximately 88°F. This ensures that the chocolate has more shine and less graininess.

A **gel** is also a dispersed system, but with rather different characteristics from the others. Its microscopic structure is somewhat similar to a sponge, with fibrous material structured in the form of a tridimensional network containing a mass of liquid within it. The fibers and liquid are united by links that are weak but sufficiently strong so that the texture of a gel is usually elastic and rubbery. Some gels are reversible and when heated become liquids again, such as gelatin mixed with water. Other gels are irreversible, such as the white part of a cooked egg, and cannot return to a liquid state.

Ice cream. A Complex Dispersed System

Many culinary preparations are very complex dispersed systems. Ice cream, for example, shows a multitude of juxtaposed phases. To start, it is a very viscous liquid that is the continuous medium in the form of a foam (because of air from beating) and a solid suspension (due to ice crystals). But there are other solid particles (sugar crystals that have precipitated due to the liquid concentrating as it freezes), calcium caseinate particles in suspension coming from the milk, and drops of fat in emulsion (fats from the milk and the egg yolk, if it has been used as an ingredient). Finally, the viscous liquid itself is a solution of water with lactose sugars, from the milk, and sucrose, which is the added sugar. To sum up, a foam/suspension with a base of emulsion, whose watery phase is a solution. Mashed potatoes are also complex: they consist of starch granules bloated with water –a gel– suspended in a liquid that is an emulsion of water and the fats from the milk and butter.

All this world, along with texturizers, can stimulate chefs to prepare new ways of presenting foods that combine dispersed systems untested up to now.

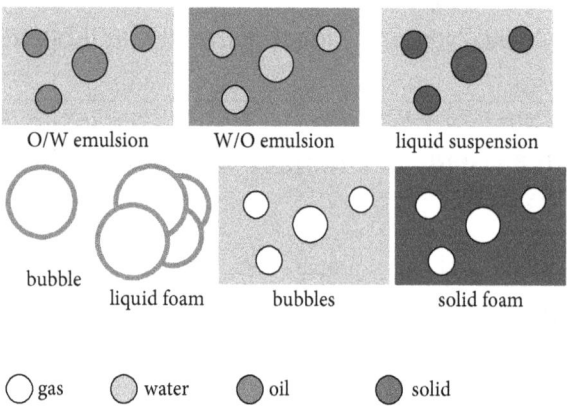

Figure 3: Dispersed systems. The top line shows diagrams of the microscopic structures of oil-in-water (O/W) emulsion, water-in-oil (W/O) emulsion and a suspension in which the particles of a solid are suspended in the middle of a liquid. The line below shows the structure of a bubble, of liquid foam, of bubbles in a liquid and of solid foam from a gas dispersed in a solid.

Chapter 7
Physical Culinary Processes

Some treatises say that traditional procedures for food preparation are the following: chopping; heating (includes cooking as a synonym of boiling, baking, frying [up to 460°F at atmospheric pressure] and pressure-cooking [up to 250°F]); cutting; drying; evaporating; fermenting; crushing; infusion; microbiological procedures; peeling; pressing; roasting-toasting; steeping; topping; cooling; distillation-rectification; emulsifying; extracting, including extraction with solvent; filtering; marinating; mixing; percolating; refrigerating-freezing; and squeezing. Other more modern and advanced processes, which will be described in

Chapter 10, should be added to these processes considered traditional.

The previous list has no logical order. There is a mixture of processes having no classification, with evident redundancies. Many are simple prepping processes, such as peeling; and others are already preparation processes, which change the physical state of foods or change their chemical composition.

This book will use a somewhat more structured classification, based on what really happens to foods when they are processed, from the physicochemical point of view. Culinary processes will be classified into three large groups:

- **Physical processes,** in which the food does not change its chemical composition but rather changes only its physical state or thermal level. That is to say, new molecules do not appear in the process, nor do previously existing molecules disappear. Some examples are freezing a mixture to make ice cream, or the preparation of a fruit smoothie by blending and mixing the fruit with milk and sugar.
- **Chemical processes.** There are chemical reactions in these, such that in the processed product there are substances initially not present in the food, and vice versa, there are compounds in the food prior to processing that will disappear. Some examples are fermentation of milk to get yogurt, or baking a chicken in the oven, which

causes changes in aroma, flavor and color due to new molecules being produced.
- **Physicochemical processes.** These are processes in which the chemical composition of substances seemingly change, but really this is not so, but rather there are only changes to their microscopic structure. An example is the denaturing of proteins due to heat, like when cooking an egg.

In this chapter the physical processes will be described, leaving the physicochemical and chemical processes for Chapter 8.

7.1. Mechanical Processes

Reduction in size is a very common need in the kitchen. Cooking with reduced-size portions has various advantages. The smaller the portion of a food, the quicker it cooks, because the heat reaches it from more surfaces that are closer to the center; this cooking is more homogenous; and the food more easily integrates and dissolves into other ingredients when mixed with them.

The **cutting** or **chopping** of a food breaks the cell membranes in plant products. Doing this frees up the liquid inside them, which facilitates thicker mixtures. Grains and seeds can thus be reduced to powder to get flours. Meats and other animal parts have a plastic and flexible texture, and their grinding produces the ground meat that, mixed with other ingredients, allows things such as cold cuts, hamburgers and meatballs to be made. In the previous list, the processes to reduce size are **chopping**, **cutting**, **grinding** and, to a certain point, **pressing** and **squeezing**. All these operations have well-known definitions. Devices to reduce size are varied, depending on the degree of reduction desired, and on the hardness and fragility of the initial food. Thus, **hand blenders**, **grinders** and **mills** have been designed for hard and brittle ingredients, **blenders** for fruits and vegetables, meat **grinders** and **slicers** for cold meats or vegetables.

Mixing ingredients is the operation that allows liquid mixtures, sauces, emulsions or doughs to be prepared. Some mixtures are already the final dish, such as sauces obtained by simple hand or mechanical mixing, for example, mayonnaise. In other cases, the mixture is a prior stage to later cooking or fermentation. To mix, **dough mixers** and **blenders** of various designs are used, depending on the characteristics of viscosity, plasticity or elasticity of the initial mixture of ingredients. Likewise, integrated devices exist such as **food processors** –commonly called *kitchen robots*– or household **bread machines**. These appliances also have a thermal energy supply that allows heating or cooking of foods directly inside them.

Some mechanical separation procedures are able to separate certain components from mixtures.

Two examples of these processes are **filtering** and **percolating**. Filtration is the procedure of having a liquid pass through a mass of solid particles, with some components from the solid dissolving in the liquid. The solid remains free of this component, which accumulates in the liquid. An example of this process is washing cooked rice with cold water when getting rid of the starch covering the grains is desired. In many cases, percolation is done with heat, and thus is a thermal process, as will be remarked on later. The preparation of coffee is an example of percolation with heat that will be seen in Chapter 10.

7.2. Thermal Processes above Room Temperature

Thermal processes in the kitchen are those that add heat to foods or take it away. Of the list of traditional methods explained previously, thermal processes are the following: heating, boiling, baking, frying, drying, evaporating, refrigerating-freezing and chilling.

The purpose of some thermal processes is to separate components. These are **drying** and **evaporation**, which eliminate water; and **distilling-rectifying**, novel process not included in the list, which separate out volatile components from a mixture (see Chapter 10). Most heating processes seek to cook foods by means of denaturation of proteins, gelatinization of starches, fusion of fats, degradation of collagen, or the chemical reactions of caramelization and the Maillard reactions. All these changes will be described in detail when analyzing chemical or physicochemical processes in Chapter 8.

In the kitchen, there are three basic mechanisms to supply heat to foods: by conduction, by convection and by radiation. **Conduction** is the mechanism by which a food is heated by direct contact with a hotter utensil. This is the case of metal grills, on which a piece of meat, fish or vegetable touches the hot plate and heats on the side that touches the iron. **Convection** is the mechanism by which heat passes to the surface of the food using a hot fluid as a vehicle. This is the case of deep-fryers, in which the hot fluid is oil, and convection ovens with hot air, or pots in which the heating fluid is the water. Finally, **radiation** is the mechanism by which a hot area emits infrared radiation that heats the surface of the food without this food being in direct contact with the hot area. This is the case of infrared cooktops to keep dishes warm before being served, oven grills, electric grills and other similar utensils. In the majority of utensils and heating systems in kitchens, two or even the three mechanisms indicated are working. Thus, the food heats from below on a grill not only by direct contact, but also because the grill heats the surrounding air, which also heats the food by convection. The same happens when grilling from a source that emanates from above, which combines radiation and convection.

Cooking on a fire's **direct flame**, quite unusual, also integrates the radiation and convection mechanisms: the center of a flame can be above 1,800ºF, and as such the contribution of the radiation mechanism is important. Also, hot gases that are a product of combustion complementarily heat the food by convection. Additionally, cooking over embers or on a barbecue, when the food is almost touching the incandescent embers, are also a combination of conduction, convection and radiation.

Induction kitchens externally look like electric hot plates, but the production of heat is owing to a totally different mechanism. An induction cooktop creates an electromagnetic field that induces some currents on the sides and bottom of the recipient that heat it, and it is the recipient heated like this that heats or cooks the food inside it. The recipient needs to be made of the correct metallic materials: in induction kitchens, utensils made of glass, ceramic, plastic, silicone or certain metals do not work. This method has the advantage of precision when cooking, and that the surface of the cooktop does not get hot.

Heating or cooking in a **microwave oven** is based on a mechanism that is totally different from the previous ones, and will be described in detail in section 9.3.

Drying and **evaporation** are operations that have the reduction or elimination of water from a certain food as their objective. The water can be moistening or superficially covering the food, and is called **unbound** water, which is relatively simple to eliminate from food. There is also other water, which is forming part of the biological structures and is inside the cells or intercellular liquids. This is the **bound** water, which is much more difficult and costly to eliminate, and which requires higher temperatures or more time.

A **concentration** is the reduction of the water content of a solution, a diluted emulsion or a suspension. It is usually done by **evaporation**, in devices appropriate for the work scale. Concentration of a sauce in a frying pan on a gas flame at home does not bear much resemblance to industrial steam evaporators used to manufacture *fruit juice concentrates* or *concentrated milk* –which is usually called *evaporated* milk– but both processes are based on the same mechanism: the transmission of heat to the liquid by means of convection.

Reduction of the amount of water in a solid is called **drying**. The classic method is exposing the substance to a dry environment and to sun, such as when drying fruits: the mechanisms of convection and radiation are combined. Hams and sausages dry in darkness and in dry and ventilated spaces, called *dryers*: the process is not based on heating the food, but rather on the diffusion of water from inside of the substance towards the dry air in the environment that circulates naturally or pushed through fans. By means of the novel technique of **lyophilization** (freeze-drying), the drying of certain foods can be done with great efficiency (see section 10.4).

Pasteurization, invented by **Louis Pasteur**, a French scientist, and the **sterilization** of foods are thermal processes whose objective is food safety. It is about keeping the food at a sufficient temperature during a certain time with the aim of denaturing the proteins from the microorganisms it contains. These can be present in the food naturally or may have entered the food due to contamination. Tinned and canned foods are pasteurized or sterilized in *autoclaves* –large pressure cookers– with steam at high temperatures. Milk and juices undergo thermal treatments at high temperatures during a brief period of time in special heaters. These are UHT (*ultra high temperature*) processes at 235ºF during one or two seconds. They give some color to the milk because of the Maillard reactions that takes place inside it (see section 8.4), and to avoid this there are thermal alternatives such as HTST (*high temperature-short time*) at 161ºF for 15 seconds. These processes are given commercial names such as *ultra pasteurization*. Alternatives to pasteurization are the **high-pressure treatment** and the **high voltage pulsed current** when the temperature of the product cannot be raised and there is no wish to add preservatives.

Cooking processes such as boiling, frying or roasting usually involve chemical changes in the product, so they will be described further on.

7.3. Thermal Processes at Low Temperatures

Thermal processes below room temperature have two distinct objectives: preservation of foods during prolonged periods, avoiding microbial decomposition and putrefaction; and the creation of certain cold preparations. All foods contain a greater or lesser amount of water, which causes the microorganisms present in the air or in the food itself to naturally proliferate in the moisture and end up colonizing and decomposing it. Many microorganisms deactivate at temperatures below 32ºF and, additionally, the lower the temperature, the lower the speed at which microorganisms reproduce.

Household **refrigerators** have different temperature levels. They are usually set at about 40-46ºF inside, and some appliances have a drawer practically at 32ºF. In the **freezer**, current appliances are at least at -0ºF, a temperature considered sufficient to freeze foods. Some models can be programmed to -22ºF. Industrial models exist capable of reaching -85ºF, always by means of the classic technology of a gas refrigeration cycle.

Alternatively, **liquid nitrogen** can be used to reach much lower temperatures. At atmospheric pressure, liquid nitrogen is at -321ºF and is capable of quickly freezing any food it comes into contact with. Cooking based on this peculiar product is described superficially in section 10.2.

Chapter 8
Chemical or Physicochemical Culinary Processes

Chemical reactions are processes in which some initial substances, called **reactants**, are converted to one or more different substances, called **products**, chemically different from the first. There are many types of chemical reactions: they can happen in a gas phase, a liquid phase or a solid phase, or in a mixture of phases. Chemical reactions that take place in live organisms are called **biochemical reactions**. There are chemical reactions that take place because of light activity, such as photosynthesis: these are **photochemical reactions**.

Some reactions take place completely, in such a way that the reactants disappear totally and in the end there are only products. In contrast, there are others in which a part of the reactants always remains in contact with the products in the final mixture, and the reaction cannot advance more: these are *equilibrium* reactions. Another aspect is the *reaction rate:* there are instantaneous reactions, like the majority of combustions, and others that are much slower, such as oxidation on the surface of fruits when in contact with air. The rate at which a chemical reaction happens usually increases as the temperature rises. Living beings secrete some protein molecules called **enzymes** that are capable of accelerating certain biochemical reactions very selectively: they are biological *catalysts*. Finally, all reactions take place with the involvement of some form of energy, normally caloric. In many cases, the start of the reaction requires an initial thermal stimulus, as in gas combustion, which requires an initial ignition point. Some of the chemical reactions in the kitchen, such as fermentations, require the action of specialized microorganisms that supply enzymes, as will be seen.

Gas combustion in the kitchen is an example of a simple gas phase chemical reaction: the reactants are the gas from the kitchen and the oxygen from the air, and the products are carbon dioxide and water vapor. In contrast, making bread is much more complex because it involves different consecutive chemical and biochemical reactions, as will be seen in a later section.

8.1. Denaturation of Proteins, a Physicochemical Process

The frontier between chemical processes, in which new molecules are produced and initial molecules disappear, and physical processes, in which there is no chemical change, seems clear but at times, not so much. We will visualize this aspect with an everyday example. The white part of an uncooked egg is made up of water and a group of proteins generically called *albumins*, which are really a complex mixture of different molecules. These proteins can be imagined coming from the linking of a long chain of smaller molecules called *amino acids*. Amino acids are connected to each other by strong links called covalent bonds. These amino acid chains are the *primary structure* of proteins. Each chain, which can be imagined as a long fiber of molecular dimensions in a watery medium, produces some weak bonds between areas of the same protein chain, causing the fiber to entwine around itself creating a tiny spring, which in turn folds over itself creating a microscopic globule. Thus, the proteins of the raw egg white can be imagined as very small globules that are independent from each other. The grouping of protein globules dispersed in water have a consistency of a viscous liquid. The spring and globule structures are called the *secondary structure* and *tertiary structure*, respectively, of the protein.

When the egg white is heated, the weak links of the secondary and tertiary structures are overcome by the thermal agitation of the medium, the fibers stop being entwined and extend through the water with only their primary structure in fibers. These can link together at some points, forming a spongy structure called a *gel*, already mentioned in Chapter 6. The water from the egg white will be retained by the protein fibers of the network. The result is a fragile elastic solid, the cooked egg white. This is the **coagulation** or **denaturation** process of proteins, which usually happens starting from 150°F.

Note that the chemical composition of a protein fiber is the same whether it is formed of a globule or if it is elongated: it is the same chain of amino acids. But the fact that the protein is folded or not gives it very different physical properties: its **configuration** is distinct. The coagulation or denaturation of the proteins changes configurations, not compositions, and for this reason this change is classified as a physicochemical change, not as a chemical reaction. In the majority of proteins, this change is irreversible. It does not take place only with an increase in temperature, but also if the pH conditions of the medium change, becoming more acidic or more basic. This has importance in certain gastronomic preparations.

Some Culinary Examples of Denaturation of Proteins

- Cook by heating the egg white and yolk from eggs, like a hard-boiled egg, an omelet or in the preparation of custard or creme brulee.
- Preparation of anchovies in vinegar, or any type of ceviche, dishes in which fish proteins denature when submerged in an acidic liquid such as vinegar, lemon juice or acidic *pickled* foods. The texture of the fish muscles modify, becoming harder, and their color changes from pink and almost transparent to opaque white.
- Preparation of a meat *tartare* follows similar steps to that of ceviche, and ground beef denatures partly because of activity from the acidic ingredients that are added (capers, lemon juice, sour cream).
- Heating milk causes coagulation of the *lactalbumin* in the milk, which deposits on the upper layer forming cream, which is not of fat but rather of protein.
- The addition of lemon juice to milk while cold coagulates another protein from it called casein, which remains inside it as granules. This is the first stage in producing *leche merengada* (a Spanish variation of a milkshake). Custards are made by mixing milk –with added cream or not– with rennet or with extracts from certain plants. These substances contain the enzyme rennin or *chymosin*, which is a protease –an enzyme specialized in attacking proteins– that was

> previously extracted from one of the stomach chambers of a calf, and which currently is also obtained through chemical or biotechnical techniques. Rennin is capable of denaturing the casein, forming a gel from which the whey –the liquid from the milk without fat or protein– is expelled by pressure. The final product is a white cheese that can later be salted, cured or fermented to get the enormous variety of cheeses that exist.

Thermal denaturation of proteins begins at temperatures above 130°F. This is the temperature that, for safety, the center of something to be cooked should reach for a few minutes. The **low-temperature cooking** currently practiced in many restaurants and household kitchens (see section 9.1) should by necessity happen above this temperature to ensure that the pathogenic microorganisms that may possibly be present in the food have been destroyed. For example, presence of the bacteria *salmonella* on egg shells can contaminate their inside, thus the practice of eating raw eggs is not recommended. Complete cooking of hard-boiled eggs or in an omelet denatures the bacteria, but eggs cooked in water, with the white part soft but coagulated and the yolk creamy, may not be sufficient to denature the bacteria, as to do this the egg must be kept at 138°F for a few minutes. Coagulation of the yolk requires 145°F, which means the margin of temperatures is limited. A method exists to get a safe egg cooked in water: the egg

–medium size– is put in boiling water for 3.5 minutes, and as such the white coagulates but the yolk does not. Then, the egg is put in a double-boiler at 140ºF for 15 minutes. The yolk then reaches 138ºF without coagulating, but without live salmonella.

In contrast, excessive cooking of hard-boiled eggs creates a greenish strip between the egg white and the yolk. This is iron sulfide, which has appeared because of the reaction from the hydrogen sulfide that the compounds with sulfur have released from the egg white, and some components with iron atoms in the yolk.

8.2. Fermentations

A **fermentation** is an *anaerobic* metabolic reaction –which takes place in the absence of oxygen– by which a component partially oxidates from the action of some microorganism, called a **ferment**. It is, then, a biochemical reaction that should be carried out at temperatures under 140ºF to avoid that the microorganisms die. The new molecules that form in the fermentations are smaller and simpler than the initial ones. Fermentation reactions have been known from time immemorial, as the preparation of bread, beer, wine and cheese are based on them. The microorganisms that carry out fermentations are of two types: **yeasts** – specialized unicellular fungi, at times forming chains– and **bacterias**, which are also unicellular organisms but without a nucleus and with a much simpler cellular structure. The fermented products industry is one of the most important in the food sector, and household fermentation methods are currently booming (Katz, 2012). Artisanal beers, bread, yogurt, pickles and all types of fermented foods are being made, which are highly appreciated by a part of the population that see them as having, in addition to their gastronomic virtues, healthy nutritional properties.

> ### Prebiotic and Probiotic Foods
>
> The healthcare system and the populace are giving greater and greater relevance to knowledge about **intestinal microbiota**, that is to say, the group of colonies of microorganisms that are in symbiosis with our organism, microbiota that is different in each individual, with similarities by geographic areas, ethnicity and family groups. For the correct maintenance of microbiota, there are nutritional trends that recommend the ingestion of **probiotic** and **prebiotic** foods. **Probiotic** foods, such as unpasteurized yogurt, contain live microorganisms that overcome the drastic acidic conditions in the stomach and reach the intestine in sufficient quantity to be able to increase the microbiota present there, or colonize it. **Prebiotic** foods, in contrast, contain specific food for microorganisms, like certain fibers, for example *inulin*, in addition to food for humans.

Fermentable substances, those on which ferments act, are those that contain some degradable or oxidizable component, and the majority of sugars from a wide range of foods are ideal candidates for fermentation. There is an enormous range of fermentation processes, although they are based on a limited number of types of biochemical reactions. The main ones are: ethanol fermentation, acetic fermentation, lactic acid fermentation and butyric acid fermentation.

Ethanol Fermentation

Ethanol fermentation transforms sugars into *ethyl alcohol* or *ethanol* by means of yeasts from the species *Saccharomyces cerevisiae*. This is the fermentation involved in the making of **wine, cider, mead** or **palm wine.**

To make grape wine, it begins with the grape juice or *must*, which contains glucose, fructose, some sucrose and many other sugars. The yeasts oxidize the sugars into carbon dioxide and ethanol at greater or lesser speed, and at variable temperatures, up to 85°F in certain red wines. In some countries, the addition of sugar to the grape must in order to increase the alcohol content of the resulting wine is accepted: this is the process of **chaptalization**, not authorized in certain regions. Sparkling wines are created by the wine going through a second fermentation, which can be done in the general vat or in the final bottle. This new fermentation requires adding new doses of sugar and yeasts. The residues produced should be decanted and extracted, and the *liqueur d'expedition* –a mixture of sugar and wine– can be added at the end, which gives it the desired degree of final sweetness.

The production of **cider** includes ethanol fermentation, and a second fermentation called malolactic, carried out by bacterias of the *Leuconostoc* genus. The process transforms the *malic acid* present in the apple juice into *lactic acid*, having lesser acidity, while also releasing carbon dioxide. This fermentation can also be induced in very acidic red wines.

Mead, which is probably the most ancient alcoholic drink, is prepared from a mixture of water with honey, which contains abundant glucose. The fermentation is carried out with *Saccharomyces* yeasts. **Palm wine** is prepared by fermentation of the sap from the date palm, which contains up to 10% sucrose. The initial fermentation is through bacterias, which turn the initial alkaline medium into acid, which allows the yeasts that produce ethanol fermentation to develop.

Obtaining **beer**, very old historically as it dates back to ancient Egypt, is based on fermentation of the mixture of starch from seeds and water with *Saccharomyces cerevisiae* yeasts and others. Any cereal, such as wheat, rice, corn, oat or barley, can be used. Not all cereals are directly fermentable, and some, such as barley, must undergo the method of *malting*. This consists in making the grains germinate: the enzymes in the cereal spontaneously act on the starches, reducing the length of their molecular chain. The malting stops, reducing the moisture in the grain and then roasting it. The degree of roasting will then determine the color of the beer obtained. Addition of hops and other ingredients flavor the drink. The subsequent fermentation transforms the sugars coming from the starches into ethanol and carbon dioxide.

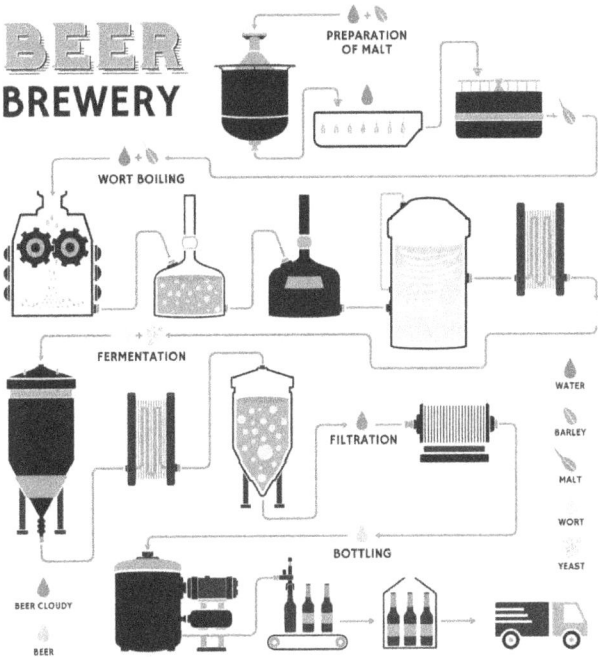

Picture 3. Beer fermentation process.

The fermentation of flours from seeds with less water than those to make beer gave rise to the preparation of the basic type of **bread**. The flour dough mixed with water and *Saccharomyces*-type yeast ferments, producing carbon dioxide as a result of the yeast breathing, making the volume of the mixture increase after 40 or 50 minutes. Small quantities of alcohol are also produced, which will be lost in the baking. The presence of *gluten* in the flour allows stretchy doughs to be obtained that nicely conserve the gas bubbles on the inside. Once the size has increased, the resulting dough is baked. At temperatures of 160°F or more, the microorganisms responsible for fermentation die, and the starch molecules degrade and link to the gluten. At typical baking temperatures, around 390°F or more, the entire dough cooks, and steam and all the alcohol and residual gas is released. At these temperatures the crust hardens, and the *Maillard reactions* (see section 8.4) produces the molecules that make it acquire its characteristic aroma, color and flavor. There are an infinite variety of seeds, types of flours and methods to make them. For example, bread can be made without yeast –*unleavened bread*– that is not spongy because gas is not formed on the inside.

Sake is the generic name of alcoholic drinks in Japan, but in our world the term refers to the drink obtained by means of yeasts from the species *Saccharomyces sake* and *Aspergillus*, along with bacterias. In a similar way to beer made from barley, first the starch from the rice must be degraded by means of enzymes and then fermentation is produced, which creates lactic acid and then ethanol.

Acetic Fermentation

Acetic fermentation transforms ethanol into *acetic acid* by means of bacterias of the *Mycoderma aceti* genus. This is a process that is aerobic, an exception in fermentations, and requires the presence of oxygen from the air. The most well-known acetic fermentation is getting **vinegar** from wine, from cider and from other alcoholic drinks: the ethanol oxidizes into acetic acid, and a range of organic compounds that modify the flavor and color of the vinegar are produced as products. *White vinegar* also exists and is obtained from the direct fermentation of ethyl alcohol, which creates a very concentrated product. Its use in the kitchen is very limited. **Balsamic** or **Modena vinegar** is a type of vinegar that is aged for years and is prepared from wine must by means of two successive fermentations, the ethanol and then the acetic.

Kombucha or **mushroom tea** is a slightly acidic drink obtained from the fermentation of a sugary tea solution, which has a small amount of residual alcohol product from the fermentation. It is a product having different flavors that can be prepared at home, even though it is marketed in various countries. Different fungi and yeasts of the *Ascomicetos* genus colonize the surface of the container and carry out the ethanol and acetic fermentations. The resulting liquid is credited with having diverse properties that are beneficial for health, although not rigorously proven. During the 1950s, many families grew this fungus, which they shared with friends and neighbors. This is a practice that continues to this day in some communities.

Lactic Acid Fermentation

Lactic acid fermentation transforms lactose into *lactic acid* by means of microorganisms from the *Lactobacillus* or *Streptococcus* genera, among others. The presence of lactic acid in the medium deactivates certain decomposition processes, making lactic acid fermentation a method to preserve foods. There are two large lines of products processed in this way: lactic derivatives and ferments from plants.

Fermentation of *milk* has existed from earliest antiquity, and in the absence of cold is an excellent preservation method for it. Lactose becomes lactic acid which, in turn, reduces the pH of the medium and is able to produce denaturation of the proteins in the milk. This turns into products, depending on the conditions and ferments, such as *yogurt*, *kefir* and different *cheeses*, with curing techniques and later fermentations depending on each recipe.

Lactic acid fermentation of *plants* in salt provides products such as *fermented cabbage* or *sauerkraut*. Pickles, carrots, cauliflower, jalapeno peppers and many other vegetables are also processed to get *pickled products*. Ordinary *olives* may also have been subject to lactic acid fermentation. Their bitter taste when they are picked unripe is due to their *oleuropein* content, which is eliminated by soaking, washing with caustic soda bleaches –sodium hydroxide– or caustic potash, that is, potassium hydroxide, and later fermentation. There are a great variety of methods for olive dressings, which are differentiated by the salt concentration, control of the pH, the degree of advancement of the fermentations and the complementary ingredients from the final bath.

Fish and certain *meat* preparations can also undergo lactic acid fermentation. *Garum*, a condiment from Roman times, is prepared by fermenting fish intestines in the presence of carbohydrates. Many sausages, such as frankfurters, pork sausage, fuet –a variety of spanish salami–, pepperoni or salami ferment with *Lactobacillus* or *Streptococcus*. The lactic acid produced gives the product a low pH, which connected with drying and the addition of preservatives such as nitrates and nitrites allow the products to be preserved without refrigeration for very long periods of time. Also, the risk that the products develop *botulinum toxin*, responsible then and now for poisonings due to unhygienic manufacturing of sausages and home-canned foods, is avoided.

Butyric Acid Fermentation

Butyric acid fermentation transforms sugars or lactic acid into *butyric acid* by means of bacterias from the *Clostridium* genus. This is an undesirable process in foods because it creates smells of putrefaction and rancidity. It can happen in spoiled wine and rancid butter. The generic *cheesy smell* –and smelly feet– is also due to this fermentation.

Other fermentations also exist that can be of interest in certain foods, such as **putrid fermentation**. This can be found in *Penicillium roqueforti*, which creates the smell and color of the green spots on Roquefort cheese and other blue cheeses.

8.3. Caramelization

A complex group of chemical reactions from the **non-enzymatic browning** group are generically called **caramelization** which, only by means of heat, decompose certain sugars, liberating brown compounds and causing changes in flavor of the initial substance. Caramelization of table sugar or sucrose is very well-known and utilized. When heated above 320°F, the sucrose initially decomposes, giving glucose and fructose, and loses a molecule of water, which causes the impression that the sugar is melting. Other secondary reactions happen that create hundreds of large-sized molecules –of up to 125 carbon atoms– having brown colors, characteristic aromas and sweet flavors distinct from those of the initial sucrose. The final product obtained is known as *caramel*, and from the physicochemical perspective is a dispersed system made up of a viscous liquid that has different suspended solids. If the heating continues, the molecules formed decompose, giving darker substances, toxic compounds and finally carbon. Other sugars also caramelize following a similar framework, at somewhat different temperatures.

There are foods with sugars in their composition that can also caramelize, for example onions. The mix of carrot, onion and celery and also other chopped vegetables is called *mirepoix* and is usually sautéed with oil, and as such the vegetables caramelize and can be served as a side dish. Certain sautés with onion and tomatoes in olive oil also caramelize.

The caramel used in cola drinks to give them a brown color is usually *sulfite ammonia caramel* (INS 150d), which is obtained by caramelizing sugar in the presence of certain salts. The product is stable in the presence of the phosphoric acid from the cola.

8.4. Maillard Reactions

This group of reactions is probably the most important in the kitchen, given that they create substances that change the color, aroma and flavor of cooked foods. These were initially discovered by the French doctor **Camille Maillard** in 1912 while analyzing the brown pigments of cooked meat. In 1953, **Hodge** researched many keys to these reactions, which still have not been completely figured out. The first stage in these reactions takes place between an amino acid and a sugar. At room temperature the reaction is very slow, but it accelerates considerably as of 250°F. This is why these reactions do not take place extensively in substances that are boiled or cooked in a microwave oven, but they do in substances that are fried, baked or roasted on a grill.

There are more than 20 amino acids in proteins from foods, and dozens of sugars capable of reacting with them. The amount of possible reactions is, then, considerable. The products from the initial reaction decompose and continue reacting with the initial materials, giving secondary reactions, in a very complex and variable framework depending on temperature and cooking time. The presence of an alkaline pH favors the reaction, as well as certain moistness in the food.

Cooking meat typically shows the Maillard reactions, which takes place between the proteins in the meat and the sugars present in its cells. In the cooking of bread or cookies, the temperature decomposes the starch from the seeds in short chain sugars, which react with the proteins from the grain's germ. The same reaction happens during the toasting of almonds or other nuts. Kitchen grills are utensils where the Maillard reactions can be perfectly observed as it develops. The hot surface of the grill can reach more than 570°F, and when a thick piece of meat is placed on it the meat undergoes a process of **searing**, which consists in eliminating the most superficial water from the piece, producing the Maillard reactions in it. The meat changes its color and its flavor, and later the meat can finish cooking on the other side until the desired level of cooking is achieved, or undergo another kitchen process such as roasting or stewing.

In the Maillard reactions, three groups of substances are produced that change the organoleptic properties of the foods: brownish colored pigments, typical of cooked meat or bread crust, called *melanoidins*; volatile substances responsible for cooking aromas; and substances that provide the typical flavors of foods cooked at high temperatures. Certain compounds are formed that contribute to better preservation of cooked foods. The nutritional value of the cooked food is somewhat reduced, as amino acid molecules have disappeared from it.

And, unfortunately, certain undesirable substances are also formed, being mutagenic and potentially carcinogenic. Thus, the presence of *acrylamide* has been detected in French fries and other toasted or fried foods. This substance, which is obtained from oil products and is of great industrial interest, is also formed as a secondary product in the kitchen from the Maillard reactions. It is basically due to the reaction of the amino acid *asparagine* with starch, both present in potatoes. This substance is potentially carcinogenic for human beings, according to still inconclusive studies. Acrylamide is also found in toasted bread or cookies. Health recommendations have been provided to minimize the formation of this substance in packaged products and in household kitchens. Specifically, the following practices are recommended:

> **Recommendations to Minimize the Formation of Acrylamide in French Fries**
>
> - Before frying, wash the pieces with hot water. This washes away some of the starch and amino acids from the surface of the food.
> - Cut the potatoes in thick slices. This minimizes the total surface of the food that is in contact with the hot oil, as the surface is the hottest point and where the most acrylamide forms. It has been proven that potato chips and shoestring potatoes have much higher concentrations of acrylamide than other forms of fried foods.
> - Follow the classic *double-frying* technique: fry initially at a temperature of around 300°F to cook the pieces inside. And, to make the surface golden and crunchy, the potatoes are fried again at 360°F for a short period of time.

Coffee has appreciable quantities of asparagine in its grains, and roasting the coffee also produces notable quantities of acrylamide. Attempts are being made to reduce this phenomenon by growing coffee plants with low levels of asparagine.

Milk contains *lysine*, which is one of the essential amino acids that the human body cannot synthesize. When *dulce de leche* -a very typical Argentinian sweet- is prepared, heating milk with sugar for a long time, the lysine and sugars create brown-colored Maillard compounds, which modify the flavor and viscosity of the mixture. The process accelerates, adding sodium bicarbonate to get alkaline conditions in the medium. It can also be prepared simply by boiling a can of condensed milk in water for a long time. The 212°F of the bath is sufficient to cause the reaction. *Toffee* for caramels or the sauce of the same name are prepared with sugar, liquid cream and butter at 300°F, and the reactions are the same. As in all kitchen recipes, the variations are infinite.

8.5. Spherifications

One of the culinary methods that gave fame to the well-known chef **Ferran Adrià**, chef at ElBulli, recognized for various years as the best restaurant in the world, were his **spherifications**, that today anybody can prepare in their kitchen as kits are sold to make them. Spherification is a simple cold chemical reaction, known for many years by the food industry, and which the aforementioned chef and his team managed to design as a molecular gastronomy dish. Two techniques exist, conceptually identical but which gastronomically lead to different preparations. They are called basic spherification and reverse spherification.

The chemical reaction in both cases takes place between a product derived from algae, called *sodium alginate* (INS 401), and *calcium chloride* (509) or another salt from any calcium. Sodium alginate is a derivative from *alginic acid*. It is a long chain carbohydrate, like starch or cellulose, that is obtained from the cell walls of *Laminaria*, a brown algae. Sodium alginate disperses in water forming a viscous colloid, which can react with calcium salts to give *calcium alginate*, which gels irreversibly. The industry took advantage of this property to manufacture gummy balls with cherry or strawberry flavors.

Spherification is one of the different techniques in **encapsulation**, by which an active substance like a medicine is wrapped in a membrane that isolates it, and which over time or by means of changes from the medium can dissolve and free the product inside. In the case of **basic spherification**, in 2003 **Ferran Adrià** and his team designed an encapsulation technique in which the reaction takes place between sodium alginate dispersed in a liquid –a fruit juice, for example– and the calcium salt from the bath. When a spoonful or a drop of juice with alginate is deposited in the bath with calcium salt, the drop or the spoonful is instantly covered in a very fine membrane of calcium alginate, which thickens over time. After some minutes, the encapsulations are taken out one by one, washed with water and the **spherifications** –the gastronomic name they were given– are ready to be served. The balls can be of many different sizes, from some millimeters in diameter –"caviar"– to some centimeters, at times called "fake raviolis." All are covered in a thin membrane of calcium alginate.

The process is delicate, as it requires precise control of the pH (not very acidic) and of the time.

In **reverse spherification**, invented in 2005, a substance with calcium salts present in its composition –such as cream from milk or a dispersion of cheese in water– is deposited by spoonfuls in a bath that contains dispersed sodium alginate. Similar encapsulations are formed, which contain the dairy substance inside them, covered with the calcium alginate membrane.

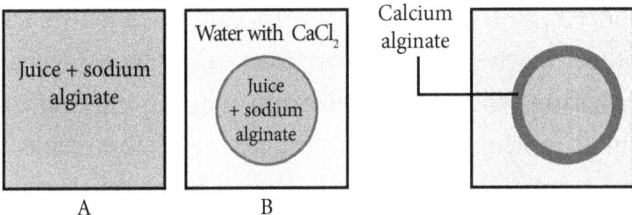

Figure 4: Outline of basic spherification. Liquid A is a mixture of sodium alginate with the juice it is desired to spherify. Bath B is a solution of calcium chloride in which a spoonful of juice has been added with alginate from A. The separating membrane of calcium alginate begins to form. Shown on the right is the spoonful of juice wrapped in the calcium alginate membrane, which once formed is rinsed in clean water to eliminate surface residue from the calcium chloride bath.

All types of products of various forms with a spherical, ball or blister shape have been prepared with these techniques: fruit or vegetable juices, whey, liquid chocolate, brandy, oil, wine, vinegar, tomato and all types of liquid or doughy substances. The pH value and viscosity are critical for the process to be successful.

8.6. Other Chemical Reactions in the Kitchen: Undesired Browning

Caramelization reactions and the Maillard reactions are two examples of **non-enzymatic pyrolytic browning reactions**, that is to say, reactions that take place at a high temperature and without action from enzymes. There are also other non-enzymatic browning reactions. Thus, in the majority of foods, heating at temperatures above 460°F breaks down the fat, protein and carbohydrate molecules, creating shorter molecules that in turn can react with each other. In this process reactions take place such as **cracking** and **polymerization** or the linking of small molecules to create bigger ones. The result is the formation of very dark, almost black substances, many of them toxic, such as *benzopyrenes*, and also carcinogenic, and at their limits decompose into carbon. These reactions should absolutely be avoided by endeavoring to not burn foods, nor toast them excessively, and to not consume the more toasted or black parts.

Enzymatic browning is due to a set of oxidation reactions of proteins and other substrates, which requires the presence of oxygen from the air, along with certain enzymes present in organisms. These reactions are responsible for the pigmentation in skin and hair, blackening of raisins or tea, and hardening of shells on crustaceans. In the case of plant foods, it is an especially undesirable process as it blackens or browns their surface when cut and exposed to air. It happens intensely in cut or peeled vegetables such as mushrooms, potatoes, apples, avocados, pears and peaches. The origin of the reaction is the presence of *chlorogenic acid* and other substrates in the fruits or vegetables that react with an enzyme from inside of the cells and oxygen from the air. In whole vegetables or fruits, the enzyme and chlorogenic acid are separated into different cellular compartments, but when these are cut, hit or peeled, some cell walls are broken down, these substances are put in contact and the reaction happens. The best treatment to avoid or reduce this is –in addition to not peeling or cutting the product until the moment it will be used or eaten– to reduce the pH of the medium, for example with lemon or orange juice, and also to reduce the temperature.

Autoxidation reactions in foods produce a rancid flavor when, once cooked, these are cooled and later re-heated. This rancid flavor is independent of what is created by the butyric acid fermentation seen in section 8.2. Unsaturated fats are responsible for autoxidations. These have a beneficial effect on the cardiovascular system but are susceptible to becoming rancid. Initial cooking at high temperatures, up to 375°F, minimize this effect, just like, in addition to facilitating the Maillard reactions, it can also produce undesirable compounds. These are effects that are in conflict with each other, as with other examples found in the kitchen.

8.7. Aging

There are many more chemical reactions that take place in foods before they are eaten. All the aging of wines, vinegars, spirits and liquors are extremely complex and relatively unknown chemical reactions. A wine, for example, contains a large amount of compounds that give it color, aroma, flavor and texture. This last property, in red wines, originates mainly due to the presence of *tannins* in the wine. Tannins are substances known since ancient times by dyers and tanners. They are *polyphenols*, a group of substances that give wine its body and astringency. *Anthocyanins* are also polyphenols, responsible for the color of red wines, which we saw in section 3.1. It has been postulated that a polyphenol from wine called *resveratrol* may play a role in reducing coronary heart disease.

The aging of wines modifies their color and reduces their astringency and bitterness. There are many other reactions inside a bottle of wine, all of them in the absence of oxygen. The aging of spirits, in contrast, is due to the interaction between the liquid and the barrels where it is stored, and which supply it precisely with tannins that give it color, flavor and astringency, something that recently distilled spirits lack.

Chapter 9
Three Kitchen Utensils: A Comparison

Each society uses the kitchen utensils that tradition and economic availability permit. The types of energy sources also put limits: if affordable energy is not available and abundant, it is not possible to have induction cooktops, microwave ovens or kitchen robots. Looking to countries that, in a rough approximation, we qualify as −Western− in the developed world, a comparison will be made below between three sets of customary utensils, focusing on the aspects most closely related to the processes that take place inside of them. This will be a sort of synthesis or integration of the previous chapters. The reference cited (Alícia-CETT, 2011) gives overall information on many of the utensils in modern kitchens.

Picture 4. Kitchen utensils.

9.1. Pots, Pressure Cookers and Vacuum Cookers

A pot, in our civilization, is usually a cylindrical container, approximately of the same height and diameter, used to boil liquid and solid foods accompanied by abundant liquid. Like all definitions, this is too limiting. A pot can be used to fry; reaching a boil is not always the desired result; and very viscous liquids, almost solids, can be cooked inside them as long as they are stirred enough. In the following we will refer to the general concept, not to uses that are unauthentic.

> **Pots that are *"avant la lettre"***
>
> Probably, the predecessors to pots were hollow spots in rocks that, once cleaned, were filled with water and the food to be cooked, and rocks that had been heated in the fire were submerged in the liquid. The evolution of the concept led to ceramic receptacles, and to metallic receptacles, and to ceramics of all types and composition, up to today. The main advantage of eating hot soups was that the food did not have live pathogenic microorganisms, which was important in times when refrigeration and food preservation techniques did not exist or were not generalized.

A boiling pot has various culinary and food safety advantages, in the first place because of boiling temperature. Pure water at sea level boils at 212°F. In the highest inhabited city on the planet, La Rinconada (Peru) at 16,700 feet, the atmospheric pressure is half the pressure at sea level, and as such water boils at around 185°F, and on the peak of Everest, at 29,000 feet, at 160°F. Thus, at whatever point on the planet, after a few minutes a boiling pot will coagulate the proteins from the pathogenic microorganisms: boiling provides food safety. Second, the temperature stays approximately constant while boiling. Thus, the cook does not have to pay attention to the boiling pot if the cooking is prolonged. Finally, cooking time does not depend on the intensity of the boiling nor the amount of food in the pot: while it is boiling, everything cooks at the same time, and it is enough to keep a minimum boil going, with the savings on fuel this represents.

The liquid and food inside the pot move for two reasons. On the one hand, before boiling, the liquid in the lower part that is in contact with the bottom, which is the hottest point of the pot, heats somewhat more than the rest of the liquid, and its density and viscosity diminish. This allows convection currents to be produced by which the colder and denser liquid in the upper part moves below, replacing the hotter liquid in the bottom, which rises. On the other hand, when it boils, the vapor bubbles produced in the entire mass pass through the liquid, mixing it up. The result is that the liquid in the pot is always more or less stirred up, which minimizes the risk of the food burning. If the substance inside the pot is very viscous, if there is very little liquid or if there are large pieces of meat or vegetables, for example, the convection currents will not move the mass inside and the evolution of the cooking will have to be watched more closely, avoiding that the food on the bottom burns. This is the case when it comes to stews.

The boiling temperature is usually somewhat higher than the boiling temperature of pure water because the preparation may have salt or other compounds, and these solutions boil at a few degrees higher than a pure liquid. For example, the water from the Mediterranean Sea, which has a high salt concentration, boils at around 217ºF. In any case, this is far from the temperatures where reactions such as caramelization and the Maillard reactions are significant. Thus, foods do not brown or toast in pots, as is well-known, unless the browning has been done before by using the pot like a frying pan, without water and only with oil or fat.

Boiling is optimal for the degradation of *collagen*. This is a protein that is omnipresent in all animal organs and tissues, and represents 25% or more of the proteins in our bodies. It could be said that the cheaper the meat, the more collagen it contains, as this is found above all in cartilage and fibers. Collagen is made up of three chains of amino acids, entwined like a triple helix, forming fibers that give resistance and elasticity to the tissues in which it is found. There are more than 20 different types of collagen, depending on the organ in which they have formed, and with some structural differences between them. However, when any type of tissue or organ that contains collagen is heated, this helical structure unravels and denatures, just like the globular proteins from the egg white denatured. Molecules from denatured collagen are sticky, and as such to cook pieces with a lot of collagen it requires time, a watery medium and a relatively gentle temperature: stews are ideal, as they take enough time to allow the collagen to denature. The process releases *gelatin*, a protein like collagen but made up of the same amino acid chains without structure. Gelatin disperses in hot water, forming a gel with a more or less liquid texture, depending on the temperature. At around 70ºF, the gel, which is transparent, sets and acquires its typical elastic texture. This process is reversible. Stews and long-cooking dishes of meat or fish extract the collagen from muscle fibers and the gelatin becomes a sauce, which thickens and sets when the temperature is lowered.

The method of first freezing octopus and squid before cooking them has the purpose of freezing the intercellular water. Given that the density of ice is less than that of water, the volume rises during freezing and can break the connections of collagen between the muscle fibers, which increases their tenderness.

Legumes are the dry seed that is separated from the sheath in leguminous plants, which contain, like all seeds, carbohydrates (up to 50% from starches, and also sugars) and proteins, from 15 to 25%, and are especially rich in the amino acid *lysine*, although their digestibility is somewhat less than with animal proteins. Some legumes such as peanuts, lupin and soy have a high proportion of fats. They all also have a large quantity of dietary fiber, up to 25% in the case of beans. To cook legumes that are dry and have little fat –lentils, chick peas, beans– requires an initial soaking for a prolonged period of time so that the inside of the grains becomes well-moistened, making their later cooking easier. While soaking, they absorb a good part of the water from the bath. The later cooking is long and mild to allow the starches to degrade.

Starches are made up of two types of chains deriving from glucose: *amylose* and *amylopectin*, this being more abundant and stickier, and which are in the form of granules in the seed. When they are heated in water above 120°F, the internal order of the granules is destroyed in a process called **gelatinization**, which in spite of its name has no relation whatsoever with gelatin. This gelatinization retains a large amount of water molecules. This softens the grain, which becomes digestible and more palatable. Glutinous rice has a high proportion of amylopectin, which is why it is more suitable for making sushi. Starches or their chemical modifications are used as food additives because of their thickening and gelling properties. Unfortunately, the molecules in amylopectin can lose water over time, which causes the food to *dry out*, such as in the case of bread, cakes or bechamel sauce. This is starch **retrogradation**, which can be reduced by increasing the proportion of fats in the recipe.

Boiling vegetables softens their cell walls, made up of cellulose and pectin, making them less crunchy, and also changes their color. It has been proposed that copper receptacles maintain the green color of vegetables, but recent studies show that salts from zinc are more efficient, and it is especially important to avoid an acidic environment in the cooking water, as in an acidic environment the chlorophyll, responsible for the green color in vegetables, transforms into *pheophytin*, having a yellowish-brown color. The use of sodium bicarbonate in the cooking water gives the water an alkaline pH, which maintains the green color and also helps to soften the cell walls. The boiling process can be done without the water touching the food, in what is called **steaming**. This is suitable for vegetables but not for legumes, which will not gelatinize. This steaming does not maintain the flavor of raw foods, as the enzymes contained in the cells are present on the surface of the vegetables when they are cut and the molecules that give the characteristic flavor to the raw vegetables degrade.

Pressure cookers derive from **autoclaves**, receptacles in which the boiling water allows the tins for canned foods to be sterilized once closed, destroying the live microorganisms that might remain inside them. Being able to regulate the pressure inside them makes it possible for the water to boil at pressures above 212°F, and the sterilization processes are faster and more efficient than in a simple double-boiler at 212°F.

Their application to cooking foods became popular in the 1920s. Current household pressure cookers work at pressures up to 20% higher than atmospheric pressure, which allows temperatures of liquid up to 265°F. At this working temperature, cooking reactions double and even triple their speed, but in essence the reactions and methods are the same as those already described for a classic pot. Some pots have electrical resistance heating, which allows more precise control of the temperatures. A disadvantage of the pressure cooker is that it is not possible to add ingredients while cooking, nor can you see the state of the food inside. Another problem is boiling products that produce a lot of foam, as in certain cases this can obstruct the steam release valve, with the risk this implies. Pressure cookers usually have a second security valve, but the risk from foam should be avoided, which is why it is not recommended to fill the pot to more than two-thirds of its capacity. In mountainous areas, pressure cookers are highly-valued cooking utensils because, as we have seen, the temperature at which water boils at high elevations is much lower than 212°F and the cooking would be very slow. With a pressure cooker, cooking times are much shorter. This is an advantage because the amount of fuel needed to cook is minimized.

With the opposite goal of slowing down cooking times and thus getting new food textures and new flavors, as of 1970 **slow cookers** (or crock pots) were developed. These are containers at atmospheric pressure that heat up by means of electrical resistance integrated into their walls and which control the desired cooking temperature, from 160 to 212°F. The times required are much longer than in classic pots, depending on the working temperature. Low-temperature cooking is similar to what is called −often inexactly− **vacuum cooking** *(sous-vide)*, in which the food is put in a hermetically-sealed plastic bag, from which the air has been extracted, and is submerged in a receptacle with water at a controlled temperature, like thermostatic baths in laboratories. In both cases, denaturation of the proteins in foods and degradation of plant walls in vegetables are produced. The vitamins also degrade less. Nevertheless, the working temperature should be above 140-145°F to ensure that possible pathogenic microorganisms are degraded during the cooking. These have the disadvantage of high electrical consumption –due to the length of the process– and that a good part of the energy is lost as external heat. The **vacuum cooker**, which is marketed under the name **Gastrovac**, really works in a vacuum. It is an electrically heated pot and is connected to a vacuum pump that lowers the pressure inside it to one-fifth of the atmospheric pressure. In these conditions, water boils at 140°F and foods can be cooked like in a slow cooker or in a *sous-vide* cooker.

9.2. The Frying Pan and the Deep Fryer

The **frying pan** is a cooking receptacle with a wide open surface and little height and is heated, like pots, by gas, electric, vitroceramic or induction cooktops. It is ideal to concentrate foods that need their water content reduced. The frying pan also allows flat foods, like tortillas or cutlets, to be **fried**. If the food is too bulky, and there is not much oil, it will only fry on one side, which is usually a problem, and often for this operation a deep fryer is preferred.

The frying pan is also ideal to **sauté** foods. This technique cooks with a small amount of oil and high heat. The food is browned superficially, avoiding that it becomes too dry. Butter is not suitable for sautéing because it decomposes at lower temperatures than other fats as it contains milk solids and water. At these working temperatures you can get caramelizations and, less frequently, Maillard reactions.

It is easy to stir the content in a frying pan, which is why it is appropriate for cooking and concentrations of thick mixtures that need manual stirring, such as tomato sauce or bechamel. In fact, the high viscosity of these products impedes the development of convection currents in the receptacle and, because of this, if it is not stirred, the parts that touch the heated bottom usually stick and burn, while the surface of the food is still cold. The cooking of these viscous foods also produces spatter, because the hotter food at the bottom can start to boil and the vapor bubbles that pass with difficulty through the mass explode when they reach the surface and part of the hot substance releases in the form of spatter. Manual or mechanical stirring is, in these cases, the only solution.

The main problem of a frying pan is the difficulty in controlling its temperature. Often, what is fried in a frying pan is cooked at temperatures that are too high, and in addition to the Maillard reactions, undesirable compounds can be produced. On the other hand, oil heated at very high temperatures can catch fire if the frying pan is on a direct flame. Oil does not come to a boil because it reaches its *smoke point* before this, which is the temperature at which oil begins to decompose. At this temperature, around 500°F, the triglycerides in the oils decompose, producing shorter hydrocarbon molecules of greater volatility. These vapors can easily catch fire, and produce accidents and kitchen fires.

> ### Advantages of the Deep Fryer
>
> The **deep fryer** solves some of the problems of frying pans. They can use liquid oil from any origin (fruits such as olives, or seeds such as those from sunflower, soy, peanut, canola, grape seeds and others) or masses to fry, solid at room temperature, such as pig fat or coconut or palm oil. The deep fryer requires a much greater amount of oil or fat than a frying pan, but control of the temperature is much better: reaching temperatures where the oil decomposes is avoided and, as it is an electric utensil, in the setting around the deep fryer there are no combustible ignition points that could ignite the vapors if they are produced. The technique of frying by immersion, in which the food is totally submerged in the hot oil, avoids the food being excessively impregnated with the oil because the rapid evaporation of the food's water from all sides delays penetration of the oil in the pores.

Frying time will depend on the size of the pieces being fried and their initial temperature. This is why frozen croquettes that are directly fried without first defrosting should be small, to ensure that the core of the piece defrosts and heats while the battered exterior does not burn. Croquettes at room temperature, in contrast, can be bigger. The shape of the piece is also decisive in estimating the required cooking time. Long or flat pieces require little time, but spherical pieces of the same weight will have to be fried for a much longer time and at much lower temperature to avoid them burning on the outside while their inside is still raw. As was seen in the section on the Maillard reactions, deep fryers have the advantage of being able to work in stages with controlled temperatures under 300ºF to ensure the inside cooks, and to avoid the formation of acrylamide and other undesirable compounds on the surface.

The main disadvantage of deep fryers is that the oil is used on various occasions, and thus ends up degrading. The reason is that the water the foods contain and what is released in the reactions from frying combine with triglycerides, proteins and sugars, producing substances that are not soluble in oil. These substances are generically known as *polar compounds*, which have one of their molecule ends being water-loving, with which it forms emulsions. Thus, the mass of oil sees its viscosity rise considerably. One of these polar compounds is *acrolein*, a volatile substance having a sharp, pungent odor that irritates the human respiratory system. The number of times an oil can be used will depend on the foods that have been fried in it, the temperature used and the type of oil, but in any case more than three or four times is not recommended. The aspect, odor and color are indicators of when the oil should be renewed. This renovation should be done completely, and small amounts of new oil should never be added to compensate the oil that has been lost in foods, because all the polar compounds will continue in the bath, increasing with each new fried food.

Other types of utensils already seen are casseroles and saucepans, intermediaries between frying pans and pots; and the **wok**, a specialized frying pan having large dimensions that allows cooking and to keep an already cooked food hot.

To avoid foods sticking to the surface of utensils, the internal surface can be covered with a fine layer of non-stick plastic, commonly called *Teflon* due to its most well-known marketing brand. This is an inert polymer based on compounds from carbon and fluoride. It is completely innocuous, even if ingested, although in certain older coatings it was accompanied by an additive –no longer in use– that could present some problems. Another solution is the use of extraordinarily smooth surfaces on a microscopic scale, such as certain ceramics or metals, that also show good anti-stick performance.

9.3. The Oven and Microwave Oven

An **oven** has an enclosed space where the material to be heated is placed, and has a heating system. In the world of industrial foods there are many types of ovens, very specialized according to the food. In household kitchens and restaurants, ovens are usually multi-purpose spaces, heated by gas or, more frequently, electricity. They have heating from the bottom of the oven and a grill on the upper part to grill or brown the top part of a food.

An oven heats foods by convection. The hot air inside the oven in turn heats the foods by circulating freely inside *–natural convection–* or by *forced convection* using an inner fan. In electric ovens, the temperature can be controlled with relative precision, from temperatures slightly above room temperature –120ºF– up to temperatures higher than in deep fryers, as an oven can reach 480ºF. This allows them to be used not only for conventional cooking at high temperatures, but also for slow cooking at low temperatures. It is enough to just put the temperature between 160 and 212ºF.

The shape and dimensions of the piece to be baked determines the temperature and baking time. The transmission of heat in the oven between the hot air and a solid food is much slower than the transmission between a liquid –hot water or hot oil– and a solid food. Although the air can be at a temperature as high as 480ºF, the inside of a somewhat thick piece to be roasted can take a few hours to reach an inner temperature of 160°F. Baking times are not proportional to the weights of the foods because they receive the heat over the surface, but the entire volume must be heated: a six-pound piece will take somewhat more than twice the cooking time as a three-pound piece of similar shape, as small objects —a small chicken— have a higher surface to volume relationship. As such, cooking thick pieces —a turkey, a ham— should be done with very long times and not very high air temperatures to avoid that the surface burns, and for thinner pieces the time should be less and the temperature higher. In any case, the surface temperature of the food can reach very high levels and thus the Maillard reactions and caramelizations will happen, which is usually the cook's objective when baking.

The upper **grill** is a metallic piece located in the upper part of the oven which, by means of electricity or gas, maintains a high temperature. It heats the surface of foods in the oven through a radiation mechanism, without direct contact, and is sufficient to brown, grill and produce the Maillard reactions on the top layer of foods. Its efficiency depends on the distance of where the oven tray is placed.

A **microwave oven** is based on physical principles that are totally different from a conventional oven. This device, invented in 1945, has an antenna inside it that emits microwaves —a non-ionizing electromagnetic radiation similar to radar— called a *magnetron*.

The microwaves produced in the magnetron are conducted to the inner cavity of the oven, where the water in the foods acts as a receiving antenna: the water molecules capture these microwaves and begin to vibrate and move, shifting on a molecular level and, as such, invisible to the naked eye. Such vibrations cause friction between molecules, which leads to heating of the biological tissues. It is, then, a radiation mechanism that induces friction. This heating mechanism is unmatched compared to the heating mechanisms seen earlier. Microwaves do not modify the food but only the temperature of its water. The oven space is perfectly closed to avoid the microwaves exiting the oven and heating the bodies of the people nearby.

Browning Foods in the Microwave Oven

In a microwave oven, the hottest point is the food. The liquid water in a glass can heat until boiling, and the inside of foods can heat to around 230 or 250°F, but not more. As such, caramelizations or the Maillard reactions cannot appear in any appreciable amount in foods heated or cooked in microwaves. Nevertheless, a method exists for microwaves to brown foods. It consists in using particular ceramic utensils, which are usually trays, having certain metals in their composition that act as a receiving

> antenna for the microwave. When this empty tray is put inside the microwave for long enough, it can heat to 350-390°F. Once hot, foods are placed on it and the device is started, so the foods brown from contact with the hot tray at the same time that they cook from the microwave. The recommendation to not put metallic pieces inside the microwave is due to the same phenomenon: metals like aluminum foil or metallic cutlery can end up getting very hot and shooting off sparks that could damage the device.

The potency of a microwave is limited by its design, and as such the cooking time in these devices is proportional to the amount of food placed inside them. The programs that microwave ovens usually have are based on making the magnetron work in more or less long intervals, taking into account the function they should carry out: defrosting a food, heating or cooking, although in some more recent models the power of the magnetron can be adjusted. Many microwave ovens also have an upper grill that works independently. Incidentally, criticisms about the safety of microwave ovens are completely unfounded, undocumented or malicious, such as those referring to the intrinsic safety of the device as well as fallacies about hypothetical modifications of foods heated in them, or analogies between microwaves and radioactivity.

There are many other utensils and devices in the kitchen, some of which integrate functions that have been described on their own earlier. **Kitchen robots** combine grinding, kneading and heating systems, which allows food to be prepped and cooked in the same device. Other specific devices are **rice cookers** or **bread bakers** to prepare and cook bread dough with the ingredients and texture desired by the user.

Chapter 10
More Devices and Methods

10.1. Coffee Makers

The objective of the **coffee maker** is to put sufficient hot water in contact with a suitably ground roasted coffee during the time necessary so that the water dissolves and extracts the soluble components that give flavor and color to the infusion. The flavor of coffee has different components, described with terms such as body, acidity, aroma, bitterness or sweetness. More than 800 chemical compounds are found in the liquid, among them coloring, aromas, vitamin B3 and others, depending on the species and variety of the coffee plant. Of the chemical compounds extracted with the coffee, the most well-known is *caffeine*, an alkaloid (see section 3.2) having a bitter taste that is also found in tea –which has been given the name of *theine*, even though they are the same compound– or in *mate*, a traditional drink in some South American countries. The amount of caffeine and other components extracted depends on many factors: the plant variety, the mixture of varieties, the degree and type of roasting, the degree of grinding and the extraction procedure.

In essence, there are two methods to prepare coffee: the percolation method, in which the hot water passes through the coffee, and the plunger method, in which the water and coffee remain for a certain time in contact, and then are later separated.

The **percolation** method is what filter coffee makers, the Italian coffee maker and capsule coffee machines or espresso machines in bars use. The difference between them is where the energy that pushes the hot water through the coffee comes from. In filter coffee makers, it is simply gravity that makes the water penetrate among the particles of the ground coffee. It requires a medium grind, and the water is usually boiling. In the *Italian coffee maker*, it is from the pressure of the steam produced in the lower receptacle where the water heats; the pressure makes the water rise through the central tube, pass through the coffee and the liquid comes out through the central tube into the upper receptacle. The pressure in this design is greater than that in the filter coffee maker. The water in this design does not reach a boil, but does reach temperatures above 190 °F. This requires a medium grind.

The highest pressure by far is found in *manual espresso machines* or those that use *capsules*. The water is pushed at a pressure 15 or 20 times greater than atmospheric pressure. This allows the hot water to pass through the capsules or portafilter that hold much more finely-ground and compacted coffee. In these coffee makers, the water that flows is usually less heated than in the other designs, at around 185ºF. The user cannot intervene in the capsules, which the manufacturer has prepared with the varieties of coffee, grinding and compacting they believe correct for the final result. *Single-serving coffee makers*, predecessors to those using capsules, work at somewhat inferior pressures, but there is a bit more coffee in each serving than with a capsule. In manual espresso machines, a finer grind is recommended than with the other designs.

The **plunger** or **piston** method is also called a *French press*, as these coffee makers were popularized in France in spite of being an Italian invention. In a glass cylinder, the ground coffee is put in contact with the hot water, where they remain mixed for a period of time. Then, the plunger with a filter presses on the mixture, the liquid passes through the filter and the dregs remain in the lower part. The temperature of the water in this method is at boiling. It requires a thick grind. In *Turkish coffee*, the granules of coffee are not separated from the liquid, they are left to settle in the cup and remain as dregs.

10.2. Cooking with Liquid Nitrogen

Since 2000, **Heston Blumenthal**, a renowned English chef, has used liquid nitrogen at atmospheric pressure and at -320°F as a preparation system for gastronomic dishes. Its use was already habitual in the food industry to freeze agricultural, meat and fish products. The extraordinarily low temperature of the product allows it to compete advantageously with classic industrial freezers, which took much longer to freeze something as they only reached around -22°F. Liquid nitrogen is not toxic and leaves no residue, but it has the disadvantage of requiring an external supply and special storage. Distinct culinary preparations have been designed with liquid nitrogen: ice creams prepared instantaneously in front of the diner, solid pearls of oil, desserts hot on the inside and frozen on the outside, and all types of novel presentations. Its use in the kitchen is reserved, for now, for chefs and restaurants serving haute cuisine.

Picture 5. Molecular cuisine. Cheesecake with chocolate cooked in liquid nitrogen.

10.3. Distillation in the Kitchen

Distillation is a normal technique in the chemical and food industries. *Spirits* and distilled liquors, of which there are thousands of types and brands, are prepared in the latter. All of them start from a product that has undergone ethanol fermentation. The distillate obtained is a concentrated mixture of alcohol and water, flavored with the volatile components of the starting product. This distillate is diluted in water until reaching the desired concentrate and, with the addition of flavors where appropriate, is often left to age for a time in treated wooden barrels that refine the flavor. Thus you get *brandy*, *cognac* and *armagnac* starting from wine or fruits, *whisky* from barley, *vodka* from potatoes and grains, *rum* from sugar cane, or *tequila* from agave. Another possibility is the **mashing** of herbs, fruits or other substrates with neutral spirits previously obtained by distillation. The mixture is later distilled, and the distillate obtained has the aroma of the mashed substrate. Sugar and water are added, and the result is a wide range of sweet *liqueurs*, such as anis -or anisette- or the different herbal liqueurs.

Since 2004, vacuum distillation devices from the laboratory, called **rotary evaporators** (commercially **Rotaval**), which are somewhat different from distillers and stills to produce spirits that work at atmospheric pressure, have entered with a certain momentum into some restaurant kitchens. Chefs distill all types of solid, liquid or mixed raw materials, and take advantage of the *distillates*, where the most volatile compounds are concentrated, and also the undistilled residue or *reduction*, which has concentrated in the less volatile products. Everything can be distilled: coffee, soups, dairy derivatives, fruit juices, vinegar, alfalfa… and the distillate and concentrated reduction, together or separately, can be taken advantage of imaginatively. It is very probable that this technique never reaches household kitchens due to the cost of the device and the dedication and space it requires.

10.4. Lyophilization (Freeze-Drying)

This operation was unknown in the kitchen but habitual in the chemical and pharmaceutical industries to prepare totally dry powdery substances. It is a procedure that does not require heating the product at high temperatures and thus avoids destruction of the active matter. Currently, it is used by businesses that prepare foods for space missions –as of the 1960s– and long-term expeditions and, more recently, for sale to the consumer. Some restaurants have the equipment and use it to prepare creative dishes.

In essence, it is drying in conditions with very low temperatures. The food to be dried is frozen by means of liquid nitrogen and is placed in a space in which a very high vacuum is created. Meanwhile, the frozen food is heated slowly by means of electrical resistors. At very low temperatures and in a vacuum, water vapor is released from the frozen solid, without it ever melting: this is the *sublimation* operation mentioned earlier (see Chapter 6). The water vapor bit by bit leaves the device until practically all the water has been released from the food. The food returns to room temperature, and now has a firm but brittle aspect, very porous because it has lost nearly all the water it contained. It can be consumed directly or submerged in a bath that permeates it with some type of distinctive flavor. These are products that can be preserved for a very long period of time. But the cost of the device, the slowness of the process and the technical difficulties in its operation make this a minority technique and, of course, unfeasible in household kitchens. The concept was used to preserve foods in mountainous areas of the Andes, taking advantage of the low night temperatures, heat from the sun and low atmospheric pressure, without the need of any device. It is now also used to prepare perennial natural flowers.

Picture 6. Lyophilization. Fresh and freeze-dried orange.

Chapter 11
Are There Limits in the Kitchen?

Of course not. Current trends are going in the direction of hyperspecialization and segmentation of the user markets. Traditional cooking, local products, slow food, vegetarian, flexitarian, vegan, halal, kosher, raw meals, macrobiotic, paleolithic and ethnic – Japanese, Chinese, Brazilian, Peruvian, Vietnamese and from as many other nationalities and regions that the reader wants, including their own– coexist, and the diner can specialize in a type of cuisine or jump from one to another without choosing any (Castells, 2016; Spence, 2017).

Products are continually being sold that up to that point were unknown in one market, but habitual in others. The reader may remember when they first saw kale, Goji berries, caramboliers, sushi and so many other products. And this is a global trend that is going to continue. In the near future, **insects** will become popular. Their consumption as human food is normal in certain cultures, but not in classical Western cooking. They are, in many cases, richer in proteins than meats and fish. The consumption of insects as food is called entomophagy, and the FDA treats insects as GRAS (Generally Recognized as Safe), so their processing, packaging and distribution is analogous to that of other species such as shrimp or lobsters. In contrast, they have been considered *novel foods* in Europe and subject to a period of observation and study. It is thought that the most valued species will be crickets, silkworms, mealworm larvae and some species of flies. The least-rejected forms of such foods will be like powdery flour, but gastronomically the entire animal is preferable. Other species with gastronomic possibilities are jellyfish, present in some Western restaurants, while freeze-dried *plankton* is on the market and some chefs are using it.

Superfoods are currently trendy, owing to the peculiar concept certain people have about what their food should be like, which comes down to being a therapy to protect yourself from diseases and achieve bodily well-being. These superfoods are nothing more than conventional foods that are said to have a special concentration of desirable nutrients, which in the case of designer foods are added by various techniques. *Chia* seeds, *spirulina* algae, pseudo-grains like *quinoa*, *kombucha*, *echinacea* or *maca* are some of these new myths. Almost all are plants, with antioxidants and normally with little energy value. The list will grow in the future, as they can be sold at notably higher prices than conventional products, and a certain population will believe them to be essential due to the nutritional advantages they supposedly add, which in the majority of cases are unjustified.

Novel appliances will also become more popular, to the extent that advertising and new fashions and trends evolve. Rice cookers, household bread makers, specialized kitchen robots have all already entered into many kitchens, just like high pressure coffee makers have as well. **Printing food in 3D** is based on the same principle as 3D plastic printers: it consists of an injection machine that is loaded with food pastes, and by means of a system of nozzles it builds the desired piece in successive layers, which can be consumed or cooked in a conventional or microwave oven.

Restaurants will specialize even more than they currently do. Trends will start appearing, such as the integration of magic into edible preparations at restaurants, or the integration of literature into menus, in the sense of preparing menus with a common theme from a specific story. As an example, it is worth mentioning proposals from **Heston Blumenthal** on his menus from *Alice in Wonderland*, his historical or thematic menus, or his multisensory cuisine, in which he integrates sound into his preparations by means of headphones with *ad hoc* sounds. The **Roca brothers** have occasionally followed this path, for example with *El somni*, a multisensory opera, or with their landscape cuisine.

> ### The Concept of Pairing
>
> The concept of **pairing** or sensory matching of dishes and drinks to each other is being worked on in the sensory sciences to detect groupings of molecules from foods that produce pleasant sensations when detected together. This would be *molecular pairing*, for which information is provided from new sub-sciences such as *computational gastronomy*, which analyzes thousands of pieces of data from global recipes and treats them with techniques from *big data*. There are some initial results from these techniques: it has been proven that there is a certain tendency for combinations of products having molecules in common to taste better, at least in Western cooking, but not so much in Eastern cooking. The field of research is immense.

Science in general, and chemistry in particular, will be capable of contributing new knowledge to this future world of cooking. And vice versa, the cooking of the future will contribute data to science so that science can interpret it, understand it and in turn create new knowledge.

Chapter 12
FAQs

This section brings together some questions that have been asked of the author at conferences or in the media. Others have already been answered in the chapters of the book. The questions are disperse and do not follow any specific order. There could be thousands of additional questions, and the author is available to respond to the best of his knowledge through email at claudimans@gmail.com.

How much iron does spinach have? Like other plants, it has a certain amount of iron: less than 0.00002%. It is a very moderate amount, and additionally it is in a form known as non-heme that the human organism absorbs very little of. The myth about iron in spinach began with a mistake in a bad transcription of a scientific article that gave it ten times more iron than it really had, and to advertising for the cartoon character of Popeye. Lentils have more iron, and red meats, clams and cockles have much more that is also more assimilable.

Why do ice cream makers always have to be in motion? An ice cream is a mixture of air, ice crystals and an emulsion of water, fats and sugars. So that ice cream is easy to lick, the ice crystals inside it must be very small, and that is achieved by continually stirring the mixture as it cools off.

Why do some fruits ripen in contact with other fruits, but the same thing does not happen to others? There are two types of fruits, *climacteric* and *non-climacteric*. The first ripen more quickly due to the ethylene gas they produce themselves in the initial phases of their ripening. Apples, peaches, apricots, tomatoes, bananas and some melons are examples of climacteric fruits. In contrast, cherries, grapes, pineapple, raspberry, strawberry, citrus fruits, cucumber and peppers are non-climacteric, that is to say, they do not produce ethylene as they ripen, nor do they ripen in the presence of ethylene in the environment. To regulate the amount of ethylene inside household refrigerators, there are devices that have an ethylene absorber integrated inside them, or absorbent devices that hang on refrigerator shelves.

What is chemical yeast? Also called *baking powder*, it is a mixture of salts that, as they dissolve in water or a moist medium or because of heat, release gases. Its composition is varied. It can be a mixture of sodium bicarbonate (baking soda) and tartaric acid that when reacting in water release carbon dioxide in bubbles, which disperse throughout the dough where the baking powder has mixed in. Another example is the mixture of sodium bicarbonate and disodium diphosphate, along with some inert ingredients. When heated in the oven, it also releases carbonic gas and water vapor, which makes the dough rise. Its actions are very rapid and substitute the biological yeast, which acts in cold because of its cellular respiration, and the carbonic and water vapor they release expand the dough outside of the oven.

Is fiber a food? Food fiber is what certain carbohydrates are called with longchain molecules, which are not digestible by the human organism as opposed to starch. Insoluble fibers, such as cellulose, do not retain water. *Soluble fibers* retain water around their molecule, which increases the volume of feces and facilitates their evacuation. Certain fibers such as alginates increase the viscosity of the mass and make absorption of the nutrients more difficult, which can help in controlling obesity. Even though it is not an assimilable nutrient, the presence of fiber in the diet is essential, which is why it figures on all the lists of essential nutrients. Around 30 grams a day is the recommended amount, better coming from foods such as legumes, grains, and fruits and vegetables.

When does yogurt go bad? As they are packaged, unpasteurized yogurts contain the mass of coagulated milk (the gel of protein and fat with retained water), along with bacterias from the fermentation and the remainder of still unfermented sugar. As time passes, the bacterias end up fermenting the residual sugar and, finally, when they no longer have food, they perish and remain in the yogurt mass. If air does not enter the pack and there are no other bacterias, the previous situation can stay for various months and the yogurt will continue being edible. As a precaution, the preferred expiration date indicated on yogurts is for around four weeks after the packaging date.

What is invert sugar? It is the product resulting from the decomposition of table sugar or sucrose to glucose and fructose by the action from an acid or from an enzyme. It is sweeter than the original sucrose.

Is palm oil harmful? It is an edible oil like any other. It has a slightly higher proportion of saturated fatty acid chains, and as such is somewhat less healthy than the oils with omega-3 fatty acids. Their cultivation produces deforestation in some tropical forests. Its color and odor are somewhat offensive for the Western palate, and the processes of discoloration and deodorizing, if done deficiently, can leave toxic residues in the oil.

What happens if a pressure cooker is suddenly opened while operating? The pressure drops brusquely and the substances inside, at 250 or

260ºF, quickly cool off and instantaneously produce a large quantity of vapor, brusquely boiling the entire contents. This drags the food, at a very high temperature, outside of the pot, putting the physical integrity of the cook at risk as they may suffer serious burns on the face and body.

There are deep fryers that claim to work without oil. Is this possible? Their design is that of a hot air oven with a fan. They cook foods at high temperatures and require more time. With pre-fried foods, that already contain some oil, they provide an acceptable final product.

What difference is there between different bottled waters? There are basically three types of bottled waters. *Mineral waters*, taken directly from the spring without any treatment, or with treatments that eliminate certain excess minerals from them. *Prepared drinkable waters* are water from the municipal network, with some additional treatment to eliminate flavor and some minerals from them. Finally, there are waters that have undergone *special treatments*, such as distillation and later condensation and addition of minerals, or the water coming from the condensation of dew. All these waters are suitable for consumption, as long as the minerals they contain are suitable for the consumer: some have a lot of sodium, and others too much fluoride for small children. The origin of the water is not important, and there is no appreciable difference between the waters in France, Italy, Fiji, Canada or the US. The presence of gas can be because the water in origin already had it, but it may have been added in the bottling. There are no appreciable differences from a nutritional point of view. It seems reasonable to use the waters that originate closer to the consumer, to avoid wasting energy in its transport.

Further Reading

The number of works edited on paper about science and cooking and science and gastronomy is starting to be considerable. Some of the ones that seem most interesting to me for the non-specialist general public are listed below. Those marked with a (c) have been cited within the text.

Alícia-CETT (2011). *Aparatos y utensilios aplicados a la cocina profesional.* Ed. Fundació Alícia & CETT, Barcelona.166 pages (c).

Alicia & elBullitaller (2006). *Modern Gastronomy A to Z. A Scientific and Gastronomic Lexicon.* CRC Press, the Culinary Institute of America. Taylor & Francis Group 2010, Boca Raton FL (USA) 247 pages (c).

Castells, Pere (2016). *La cocina del futuro.* Tibidabo Ediciones, Barcelona. 264 pages (c).

Fullerton-Smith, Jill (2007). *The Truth About Food.* Bloomsbury Publ., London. 240 pages.

Katz, Sandor Ellix (2012). *The Art of Fermentation.* Chelsea Green Publ., Vermont, USA. 528 pages (c).

Mans, Claudi (2010 y 2014). *Sferificaciones y macarrones.* Ariel (Planeta), Barcelona. 307 pages.

Mcgee, Harold (2004). *On Food and Cooking.* Scribner-Simon & Schuster, New York. 884 pages (c).

Mcgee, Harold (2010). *Keys to Good Cooking.* Penguin, New York. 552 pages.

Segnit, Niki (2010). *The Flavor Thesaurus.* Bloomsbury USA, New York. 383 pages (c).

Shepherd, Gordon M. (2012). *Neurogastronomy.* Columbia University Press, New York. 267 pages (c).

Spence, Charles (2017). *Gastrophysics: The New Science of Eating.* Viking-Penguin, New York. 493 pages(c).

This, Hervé (2012). *La cuisine note à note.* Belin, Paris. 208 pages. The author has many other works on the same subject.

This, Hervé; Castells, Pere; Kurti, Nicholas; Mans, Claudi (2017). "*Ciencia y gastronomía. Diálogo, tradición e innovación*". Investigación y Ciencia Temas nº 89. Collection of articles. 95 pages (c).

www.ingramcontent.com/pod-product-compliance
Lightning Source LLC
Chambersburg PA
CBHW061647040426
42446CB00010B/1623